AF316830

Without Our Consent

BOB SKERSTONAS

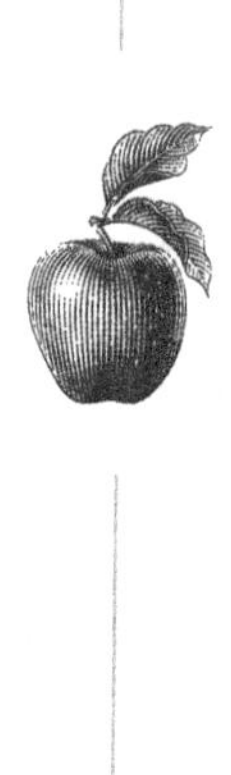

Without Our Consent

Contents

Prolog

The purpose of this book is to share with the world the effects that pesticides (defined as insecticides, herbicides, rodenticides, bactericides, fungicides, and larvicides) can have on people in their diets and lives.

Many Americans are affected by pesticides being applied to crops they eat. About three decades ago, scientists learned that even low doses of pesticides and other synthetic chemicals can harm children with exposure to chemical mixtures. 1 In another article, one in three people across America have detectable levels of a toxic herbicide linked to cancers, birth defects and hormonal imbalances. There are about 600 US agricultural and residential products containing chemicals which can be ingested through the skin, mouth, and nose.[2]

News flash – 90% of the US population in a 2018 study have detectable concentrations of pesticides or their metabolites in their urine or blood samples this is from the National Library of Science. [3]

Now for my personal story about how I am affected by pesticides in our food system.

As of this writing I am semi-retired. This is significant because I grew up on a farm in Connecticut. We had a family garden next to the Northfield Pond, a pond about eight acres in size. This was my playground until about age 12 when we moved closer to grandma's farm.

In the spring, my father would drive the farm tractor from grandmas to our house and till the land to ready the garden for planting. When the garden was planted, we had corn, potatoes, tomatoes, carrots, cucumbers, green beans, squash, and pumpkins for Halloween. The fun was picking the vegetables when they were ready to eat.

---

1 https://www.thenation.com/article/archive/pesticides-farmworkers-agriculture/
2 https://www.theguardian.com/environment/2022/feb/09/toxic-herbicide-exposure-study-2-4-d
3 https://www.ncbi.nlm.nih.gov/pmc/articles/PMC5814112/

One thing is for certain, I did not grow up with pesticides being applied on my vegetables or fruit or grain fed to animals. Fresh from my grandparents having a farm, I was able to eat everything including whole milk, beef, chicken, eggs, broccoli, corn, tomatoes, carrots, beans, potatoes, asparagus, pumpkins, blueberries, apples, pears and more.

While growing up, if there was a worm in the corn it was cut out. Rarely did I see more than a couple of non-edible ears of corn or tomatoes with bugs. When you pick apples occasionally you will get a worm in the apple. You know what you do when you find a worm in an apple? Cut it out and give it to your friend. LOL!

On the farm property, grandma had her garden. We had our garden, aunts and uncles had their gardens as well. My father most likely used some boric acid on the tomatoes as he loved to have the biggest beefsteak tomato crop otherwise no pesticides were being used. When these steak tomatoes were sliced, they were bigger than a hamburger bun. A big juicy tomato, not those little tomatoes we see in the markets today!

On the farm we also had a couple of dairy cows, and I was fortunate enough to grow up drinking raw milk. I remember feeding molasses and oats to the cows, as this sweetens the milk. We had cream on the top, and you had to shake the glass bottle of mike so the cream would mix into the milk. It was delicious milk!

As the farmland and cows were sold off, we bought milk from the grocery store. Little did I know the milk pasteurization process used chemicals to keep milk on the shelf without going sour. Today I cannot drink milk contents because I am allergic to milk or to the additives in milk and now, I drink almond milk instead.

Grandma had chickens that provided fresh eggs and fed them with grains. Farm grains in the 1960ies may have had pesticides but not to the level they are added to feed today. Our farm fresh eggs yoks had this incredible bright orange color. Today when we buy eggs at the market, even organic eggs, most of the time the yok color is still not that bright orange color as the farm fresh eggs were long ago.

During these young years of my life, we did not use pesticides on the farm. There were no bags or bottles of pesticides in the barn or sheds.

We are going to jump forward to 2016 when the mysterious rashes of unknown origin started.

Please note Without Our Consent has 399 footnotes that support information contained in this book. You may download Without Our Consent as an eBook.

About the author

Bob Skerstonas as an author has written two previous books *Best Little Real Estate Investment Book*, and *The 4th Revolution: Automation, Robots and AI*. His new book *Without Our Consent* is about a personal experience and finding out he is allergic to pesticides. He has spent over a thousand hours researching pesticides and how they are affecting all living things on our planet. There are many possibilities when researching pesticides and how they affect all living things. Bob found that there are solutions to the ubiquitous use of pesticides, and we can grow all of our food without poisoning the planet.

Cause of chronic rash and allergies

In 2016, I started to develop a skin rash on my feet and legs. I used all kinds of over-the-counter creams like Benadryl to relieve itching and get rid of the rash and Aloe Vera to refresh the skin. Imagine your body getting a rash starting at your feet and then moving up your legs. The inflammation is red and itches like poison ivy (actually looked like poison ivy). I had no idea what could be causing this skin rash. Over the next 2 and a half years, the rashes became progressively worse. No end in sight, and no known cause!

First, I started to blame nature for the itching on my feet and legs. We would go hiking at a nearby park 3 or 4 times a week. When I returned home my feet would start to itch. Blaming nature made some sense as a possible reason. I was looking for sand fleas biting my feet and legs. When you do not know what is causing an itch and rash your mind is thinking of so many different possibilities. The process of trying to figure out what was causing itching in my feet and legs went on for nearly two and a half years without a solution.

At the same time, my wife was not having any similar symptoms to my inflammation attacking my body. This eliminated sand fleas or other types of bugs attacking me on the hikes. Rashes can be caused by sensitivity to clothes, beauty products, plants, chemicals, or food allergies. There are many different types of skin rashes and can vary in appearance tremendously, and there are many potential causes. [4] [5] [6] However, I never saw a medical doctor that had an educated guess or a treatment that worked.

---

4 https://my.clevelandclinic.org/health/diseases/17413-rashes-red-skin
5 https://www.healthline.com/health/rashes#Shingles
6 https://www.medicalnewstoday.com/articles/317999

Initially I visited my primary physician here in Santa Clarita, California. He did not have a definitive diagnosis or what the cause might be. He touched the inflammation and referred me to a hematologist and a dermatologist.

The hematologist ran a battery of blood tests. There was nothing in the blood tests to report. Then a dermatologist performed an initial skin allergy test where they tested forty different possibilities of skin reactions. This first test did not yield anything except an allergy to dust mites and ragweed, which would not cause my rashes.

After the first doctors' visits, I located other dermatologists who could possibly make a medical analysis of my inflammation. I continued making appointments with new dermatologists to see if any doctor could make a diagnosis. Each dermatologist started with a standard scratch test on the skin and they continued to try and ascertain the cause of my rash. After going to nine dermatologists in all, I had 540 different skin tests! These tests reported allergies to some wild grasses, pollen, and dust mites, none of which caused my rashes. Three of these dermatologists prescribed creams that literally peeled off my skin along with other horrible side effects. [7] All medications prescribed did not work, and many made my skin condition worse! Seeing all these medical doctors was an exercise in futility. Another half year rolled by with no relief in sight.

This antidote is about one dermatologist I went to visit. In the first appointment she looked at my legs and feet for my exam. The doctor never touched my skin, not even with gloves on her hands. She prescribed some medication and off I went. A week later my itching and rash had not changed. I called the office stating the itching is not going away. Upon returning to her office and asked if she had other ideas as to what might work to stop the rash and itching. She told me to leave and never return.

In 2017 the rash became even worse if that was possible. I was taking Benadryl tablets to the point of being dangerous to

---

7 https://my.clevelandclinic.org/health/diseases/17413-rashes-red-skin

my health. Basically, I had given up on the dermatologists as their remedies were far worse than the disease.

My quest continued to find a dermatologist who could treat my rash and itching. Living in Southern CA., I was able to locate dermatologists at USC and UCLA. Using the Internet, I located the so-called best dermatologists in the Los Angeles area, who had PhDs and taught at prestigious universities. One doctor had me undress and I stood naked for a good 3 to 4 minutes while he looked at my body. At this time, my legs looked like they had leprosy. The sight of my legs was very ugly indeed. Actually, the sight of my legs was so scary because no doctor had an answer and none of them would touch my skin.

I continued to make phone calls to find a dermatologist who could possibly help me with my rash and itching. None of the above-mentioned physicians were able to help but it did cost me a few thousand dollars in drugs that were useless.

During this time, on Friday evening friends would meet at a hotel and share cocktails, a little liar's poker and discuss the week's events. We always solved all the world's problems each Friday. However, before going to this weekly reunion, I would stop and grab a burger at a local fast-food joint before going to the hotel. After I ate these yummy burgers, I would start to itch on the way to the hotel. Not a rash, just an annoying itch on my face and head. Continuing this tradition of buying the burgers on Friday lasted for several months until I realized eating the burgers was causing the itching sensation. I also developed a small rash on my ring finger on my left hand that would itch after eating the burger. It does itch even today when I eat non-organic food. At that time, I still had not figured out what was causing the itching and neither had the dermatologists. This will continue until a meeting with the Chinese doctor.

With each new dermatologist, I had to explain what the previous doctors had tried. Two of the nine doctors actually touch my skin. With no relief in sight, I was getting desperate. There seemed to be no solutions from medical doctors. Finally, in desperation I told my friends I would go to Thailand to try and get rid of this rash. Our friend who knew I was in a living nightmare, suggested a Chinese doctor who specialized in herbs. Through a friend of a friend, we located is a Chinese doctor in San Gabriel CA to give the college try before going to Thailand.

Finally, I met with the Chinese herbal specialist. I went back to a small room and took off my pants for the doctor to see the rash. He actually reaches out and touches my leg in amazement. Then he looks at me and says, 'I'll have you cured in 24 hours." The first thing I thought of without actually saying it was "Does this guy carry his balls around in a wheelbarrow." Of course, I did not say this crass phrase, but I certainly thought it. Next, the doctor and I went out to the front waiting room where my wife and her friend were waiting. The doctor then states to this room full of people: "He will be cured in 48 hours." Now you know what I was thinking, this guy really does carry his balls around in a wheelbarrow.

The Chinese specialist prescribed a list of food/drinks not to ingest, plus an herb to mix in water and a capsule to take. After nearly 30 months dealing with the American medical system, I actually received a magical reprieve from a Chinese herb specialist.

He asked his support staff to get me some herbal power and mixed this into a small paper cup for me to drink. Then he presented me with a list of foods I could eat plus dos and don'ts.

- *Keep skin moist.*
- *Avoid excessive sweating, exercising, or swimming.*
- *Do not eat hot, spicy, or fried food: Garlic, black pepper, pepper, chili, beef, lamb, duck, peanuts, nuts, half cooked egg, fried chicken, or seafood parentheses only steam fish and shrimp are OK to eat).*
- *Try not to take drinks that can cause body to heat up (e.g., Tea, coffee, alcohol-based drinks, or coke).*
- *Do not eat certain tropical fruits (e.g., Mango, Longan, Durian, etc.)*

- *Try not to drink milk, drink soy milk instead of dairy products. (I drank almond milk)*
- *Do not scratch irritated skin (use a wet towel or ice pad to relieve the itching.)*
- *Do not take "hot" water showers or bath.*
- *Do not stay in the sun too long and avoid strong sunlight.*
- *Take proper medication to alleviate itching (e.g., herbal medicine or applying cream to skin.)*

My new diet was going to be pretty limited – salmon, chicken, eggs, cereal, asparagus, broccoli, carrots, potatoes, mushrooms, green tea, almond milk. In the past. Before the rash appeared, I could enjoy liquor: scotch, beer, vodka, or red wine. We would go out to breakfast, lunch, or dinner at Thai, Japanese, Mediterranean, Italian, or Greek restaurants. Not realizing these restaurants was also contributing to my allergic reaction to food.

Even when on the herbalist diet, I discovered that some foods were giving me an allergic reaction. We had dinner one night with canned corn. Within two hours I developed a rash that didn't go away for several days. We gave the remaining canned corn away to a friend whose family could eat it.

Another incident a few months later: there was a plate of apple slices all laid out. They looked inviting and I ate one slice of apple and within one hour there was a rash on my ankle. These were not organic apples and I have not had apples at home since.

After this allergic reaction to the corn and apple, I became more vigilant and limited my diet to organic foods. I found that when I was very strict, and ate only organic food, I did not get a rash! When I go out to eat, I can rarely find organic food on the menu. So, any time I go out to a restaurant, I get a rash!! And what is the main difference between organic and non-organic food? Organic food CANNOT be sprayed or raised with pesticides!

After reading this, it should scare the heck out of you about the allergic reaction I have to food here in the United States. Yes, this is an unscientific experiment because it is one person's discovery. Also, some dermatologists had heard from several people with an undetermined rash, and they did not have answers for their patients as well. Just an observation, in my opinion, there is something drastically wrong with so many pesticides in America's food system.

Using all this information above, it took me over three years to finally figure out I was allergic to food that contains some type of pesticides. There was absolutely no help in the beginning from doctors and I was on my own to find out how to solve the problem.

After contacting the Chinese herb specialist who was able to help me get my life back, I wrote a letter to the CEOs of major food corporations like Vons, Ralphs, Costco etc….. The following is the letter.

Dear CEO,

I am writing to your company about pesticides in food. About three years [March 2019] ago I started having a skin reaction or rash. Initially we considered it was bugs biting my body and was causing the rash. After eighteen months and going to nine physicians, I had test after test, and they did not have a clue. Basically, giving up on western doctors, a friend recommended a Chinese's doctor who had a solution. The Chinese doctor restricted my diet to salmon, organic chicken, organic eggs, homemade bread, and organic vegetables. With the restricted diet and within 48 hours, the rash was gone.

Going out to eat is a giant challenge and if I go off my diet even a little the rash comes back. I cannot eat even a plain burger without receiving a rash.

Bread must be 100% organic whole wheat, or I'll face the consequences. Corn and soybeans are troublesome because companies are hybridizing corn and soybeans to be resistant to Round Up and Dicamba. These chemicals are in our food system. As an example, the breakfast cereal Honey Nut Cheerios is now off my list of food to eat. The past two years have been tough trying to find foods here in America to eat that are not treated with chemicals.

Today chemicals sprayed on crops include petroleum and mineral oils for insecticides, copper for fungicides, sulfur for fungicide/insecticides, glyphosate, Banvel, Diablo, Oracle and Vanquish for herbicides, and aluminum phosphide for fumigants. [8]

There are dozens of organic products on the market today to prevent insects harming crops. There are also dozens of organic products to spray for weeds.

Finally, you need to ask yourself: Do I want my children, friends, or relatives to experience unknown allergies or other illnesses caused by chemicals in our food system? Are these pesticides killing the bee population? If you care, then ensure your products are not part of the problem. If you are going to do something, please let me know!

Sincerely,

I received a positive response from Costco by a letter, Vons called, and Kroger sent an email reply.

Questions you need to ask yourself. Do I want my children, friends, or relatives to experience unknown allergies? What other unknown effects are these chemicals having on everyone's body and environment? What solutions can be implemented and quickly? Are these chemicals killing the bee population? Basically, all food that is not organic has been applied with chemicals. If you care, then please act!

Whether this letter had an effect I'm not sure however, over the past couple of years there has been an increase in organic products being offered for sale at these food stores. It is really a shame consumers need to ask for organic food products to be made more available.

---

8 https://www.cdpr.ca.gov/docs/pur/pur15rep/15sum.htm.

Allergies

Even with this herbal treatment, there was still a chronic itch and rash if I did not stick to the strict diet. I continued my research. I found the website www.testmyallergy.com and sent a hair sample to them. About two weeks later a list of allergies was returned. This allergy list contained fabaceae family, poaceae family, rosaceae family, asteraceae family, tea families, asteraceae family, and dairy. It seems very bizarre that after living this long I could be suddenly allergic to this many food items plus leather and cotton. This just did not make sense. However, these results changed my life! My itching became under control so long as I stayed on this even more limited food diet. I also had to eliminate wearing cotton and leather.

Using testmyallergy.com was a wonderful discovery that made me realize that my allergies are much greater than I perceived they were. The analysis showed that I was allergic to cotton. I cannot wear cotton clothes or sleep on cotton sheets. However, cotton towels are okay apparently because I only use the towels for a short period of time.

Wearing blue jeans made of denim was contributing to my rash and itching. In the beginning of this rash episode, I thought I was being bitten by bugs while going on a hike in the park. No, the allergic reaction was caused by cotton socks. Cotton socks and jeans were causing the itching and what looked like bug bites on my legs, feet, and ankles.

In the past, I would wear cotton denim blue jeans as a basic attire because it is southern California. Wearing short sleeve shirts made of lightweight synthetic material. It is important to note I also wear synthetic boxers' shorts as well. All this is important to note because the rash coming up my legs stopped at the bottom of the boxers. No cotton was touching my body in this area, and this is where the rash ended. What was really fascinating was that none of these dermatologists realized the correlation.

What I did next was to stop wearing cotton clothing that could touch my skin. I had hiking pants made of polyester material and started wearing these immediately. I bought socks made out of bamboo material and within a couple of weeks my remaining itching on my legs and feet would diminish substantially and within a month the rash was gone.

Testmyallergy.com was a key ingredient that helped change my life. I learned I could not drink any beer, wine, or liquor where the ingredients were grown using pesticides. Why? Because beer, wine or liquor can contain pesticides in the process of being produced. Now there are organic products of beer, wine or liquor that are safe for me to drink.

More to the wine story: as stated I like drinking a good bottle of wine. My buddy and I really enjoyed drinking wine. As the skin rash allergy subsided, I became more embolden to drink red wine. We would have dinner and he and I would drink two bottles of wine and on occasion three bottles of wine. The problem I have since learned is drinking too much alcohol can increase your chances of high blood pressure, which is a risk factor for kidney disease as well.

Unfortunately, I was enjoying the wine too much, and my blood pressure jumped to a very high number 180/89 taken with a blood pressure monitor at home. I went to the primary physician the next day and he informed me drinking too much wine can cause a spike in blood pressure. He told me about how the kidneys cannot process alcohol which can raise your blood pressure.

Now I only drink red wine occasionally and only drink imported beer and rums, as these two seem to have no effect on my blood pressure. My blood pressure remains around 120/70. The only time my blood pressure goes over this is when I eat out and the food is not organic, then my blood pressure can go to 140/80.

Multiple chemicals are making it into the food system. A long list of foods that I cannot consume for fear of receiving an allergic skin reaction. I cannot eat processed meats as pesticides are applied onto grains the animals eat.

Limited diet

While sharing my food allergy story with friends and acquaintances, several of them shared their stories of going to Europe and being able to eat food there without having a reaction like the experience here in America. What is the primary difference between food in Europe and the US? The use of pesticides!

By staying on a very strict organic diet, I do not get the painful rashes. Today my diet here in the US consists of salmon, organic chicken, tortillas and organic chicken sausage, organic blueberries, from Costco, Organic pizza and eggs and bread from Trader Joes, organic vegetables, orange juice, butter, almond milk, organic pomegranate juice and organic flour from Azure. As you can see my choice of food is limited to about a meager dozen plus items.

Being creative there are recipes that can be used to make breakfast, lunch and dinner. However, these recipes do get old, and I have several creative desserts I can make with organic flour. (See Appendix recipes) for some recipes that might help others who are having allergic reactions to food. - bread, cake, blueberry muffins, chicken recipes, salmon recipes see the recipes in the appendix.

Unfortunately, where I live there are few organic restaurants. I feel anxious about eating out when the restaurant does not have organic food. When I eat out and the food is not organic, I have an allergic reaction and my blood pressure usually increases. An allergic reaction can last for a day or several days depending on the food eaten.

Spices used on our food to add flavor can also be treated with pesticides, even common items like pepper and paprika. Most of us are social creatures; we enjoy going out with friends, eating, and having drinks. In my case, as with others reading this, it can be a challenge because there are a limited number of restaurants to visit serving organic and/or non-GMO foods. In the past I took going out to a restaurant for granted. Friends would make a recommendation to meet at any restaurant and we would go there and eat. Today I tried to review the menu online to review foods I can eat without getting a rash. This places a damper on going out to eat with friends.

The past few years have been tough trying to only eat organic food. One would think if you lived in California there would be organic restaurants, but the truth is there are not that many. Going out with friends over these past several years has been an absolute challenge. Because friends ask, "Can you eat at this restaurant and sometimes, I just give up and say let's go and eat. In reality though when I do eat at restaurants and it's not organic, then I usually end up with a skin rash within a few hours. My most successful alternative to going out to eat is seafood. I am still careful, if seafood is seasoned with non-organic spices, it will cause an allergic skin reaction.

Today active ingredients applied on crops include petroleum and mineral oils for insecticides, copper for fungicides, sulfur for fungicide/insecticides, glyphosate, Banvel, Diablo, Oracle and Vanquish for herbicides, and aluminum phosphide for fumigants. [9]

Rachel Carson wrote Silent Spring 60 years ago [10] and she talks about pesticides in our food system.

"Hundreds of thousands of dollars were spent by the chemical industry in an attempt to discredit the Silent Spring book and to discredit the author." [11]

From Rachel Carson time as of writing was based on the 1940ies and a proliferation of pesticides being introduced without oversight as there was no environmental protection agency.

My purpose is to get your attention about pesticides in our food system from farm to fork and how these pesticides can affect our daily lives.

9 https://www.cdpr.ca.gov/docs/pur/pur15rep/15sum.htm
10 written by Rachel Carson in 1962 Silent Spring
11 https://massivesci.com/articles/science-heroes-rachel-carson-ddt-silent-spring/

A cross over moment

The "ah-ha" moment came after visiting my Chinese doctor, and testmyallergies.com report and our trip to Vancouver, Canada. Why, because all of these combined reports and personal analysis became part of my knowledgebase to create a manageable working solution.

Next it is possible that my food allergies started in the 2016 timeframe due to various factors such as changes in my diet or could there be an exposure to new allergens, or changes in my immune system. Because it is possible for individuals to develop allergies later in life, even to multiple food items.

I was the only person, a single anomaly to be allergic to this many food items as well as other items that are in the allergy report. Doubtful because many people have multiple food allergies or sensitivities. Additionally, the specific combination of allergies can vary greatly from person to person.

Was Rachel Carson book Silent Spring correct in predicting what would happen to us by consuming food treated with pesticides? Rachel Carson's book Silent Spring is widely considered a groundbreaking work that highlighted the environmental and health risks associated with the use of pesticides. While the book focused on the negative impacts of pesticides on the environment, there is some evidence to suggest that exposure to pesticides may also be a risk factor for the development of food allergies. However, more research is needed to fully understand this risk of consuming food treated with pesticides.

Pesticides in our food system

Over the last 50 years, pesticides have made it into our food system. Basically, pesticides have become ubiquitous in the US food system, and this is a problem because pesticides have harmful effects on living organisms. Today almost all commercial food products grow in the United States agricultural system have one of 1,000 possible chemical pesticides applied in our food system. "According to data from the USDA's Pesticide Data Program (PDP), over 99% of samples of fruits and vegetables tested in 2019 had detectable levels of at least one pesticide residue." [12]

Today there is 1.1 billion pounds of pesticides applied to crops in the United States annually and approximately 5.6 billion pounds are used worldwide. [13] There are multiple types of pesticide to rid of the crops of insect using insecticides to eliminate insects, herbicides on plant species, elimination of rodents using rodenticides, removing bacteria using bactericides, controlling fungus using fungicides, and getting rid of larvae using larvicides. [14] Why are so many pesticides used on crops? Because the pesticide industry is a 77.5-billion-dollar industry and expected to grow at a compound annual growth rate (CAGR) of 5.7% from 2020 to 2027. [15] The opportunity profit is obviously the motivation to chemical companies to sell pesticides to farmers. [16]

There is a lot of data that supports pesticides and the horrible effects on all living organisms. A body of work published between 2007 and 2016 documented negative impacts of pesticides on human health and our environment. These studies found pesticides had increased "risk of cancer, neurological disorders, reproductive

---

12 https://www.ams.usda.gov/sites/default/files/media/PDP%20FY%202019%20Annual%20Summary.pdf

13 https://www.epa.gov/sites/production/files/2017-01/documents/pesticides-industry-sales-usage-2016_0.pdf

14 https://www.epa.gov/pesticides/types-pesticides

15 https://www.grandviewresearch.com/industry-analysis/pesticides-market

16 https://www.who.int/news-room/fact-sheets/detail/pesticide-residues-in-food

harm, and developmental problems in children." [17] We cannot take for granite agencies like the environmental protection agency (EPA), U.S. Department of Agriculture (USDA), or Food and Drug Administration (FDA) are looking out for are best interests in protecting people against the onslaught of chemical companies in an industry with revenues exceeding 77.5 billion dollars every year and growing.

Now it is 2023 and I am writing about my experience and the clues that consuming foods with pesticides give me an allergic reaction. Yes, I have allergic reactions from eating food that has been treated with pesticides. However, when I consume organic foods, I do not have an allergic reaction. To support this evidence a recent study on organic foods established lower levels of pesticide residues compared to commonly grown foods. [18]

Other studies report that there is a small difference between organic food and food treated with pesticides. Organic farming restricts the use of synthetic pesticides, so organic foods generally have lower levels of pesticide residues. This can be particularly important for individuals who are more susceptible to the adverse effects of pesticides, such as pregnant women, infants, and people with compromised immune systems. [19]

In this latest study from 2019 there are 35 reports meeting the criteria for review. Too few trials met the health outcomes associated with organic food consumption. Studies analyzed differences in pesticide exposure or different criteria. Positive results were observed from eating organic food and showed reduced incidences in the following:

· *infertility*

· *birth defects*

· *allergic reactions*

· otitis media

· *pre-eclampsia*

· *metabolic syndrome*

· *high BMI*

· *non-Hodgkin lymphoma.*

---

17 https://scirp.org/reference/referencespapers.aspx?referenceid=3125313
18 https://pubmed.ncbi.nlm.nih.gov/22944875/
19 https://ehjournal.biomedcentral.com/articles/10.1186/s12940-017-0315-4

Consuming organic food does not provide conclusive evidence that eating organic food provides safety from eating food treated with pesticides. However, there are a growing number of findings linking health benefits to eating organic food. [20]

Traveling outside the United States

Having spent a lot of time in Thailand, we planned a long trip in 2017. The objective was to see if my allergic reactions to food would occur while in Thailand. On the second day I started my food experiment by eating bacon, the third day sausage, the fourth day homemade pasties, all without reaction. Basically, by the end of our 30 days in Thailand I was able to eat anything without receiving a rash or itch (my wife was impressed). Compare this experience to not being able to eat so many varieties of food in America.

On another visit to Thailand, within a week the rash I had on arrival went away. After returning to the USA the rash was back in less than 24 hours.

This is my theory about pesticides in the food system, in and outside the US. Traveling to Thailand for business trips from 2004 to present day, I noticed my rosacea would completely go away from my face. Noticing my face clearing up while visiting Thailand was fascinating and I did not realize what the explanation could be. It took several more years for me to realize food in Thailand was healthier for me to eat than here in the United States. Each trip going to Thailand reinforced my thoughts as the food seemed to be better for me.

Researching this notion, I found Thailand has been promoting sustainable farming practices for several decades. The country has a long history of agriculture, and the government has recog-

---

20 https://www.ncbi.nlm.nih.gov/pmc/articles/PMC7019963/

nized the need to balance agricultural development with environmental sustainability and the health of its people.

In the 1980s, the Thai government introduced a national policy called the "Sufficiency Economy Philosophy" (SEP) [21] to promote sustainable development and improve the livelihoods of small farmers. The SEP emphasizes the importance of self-reliance, moderation, and resilience, and encourages farmers to adopt sustainable farming practices that conserve natural resources and reduce the use of chemical inputs. Additionally, there are programs promoting sustainable farming practices that can reduce the reliance on pesticides altogether, such as integrated pest management (IPM) and organic farming methods. [22]

A lesson learned is small farm producers of fruits and vegetables are much less likely to use pesticides. There are a lot more small-scale farmers in Thailand and food producers tend to use fewer pesticides. As in the United States, small scale farmers adopt more sustainable and environmentally friendly practices compared to large-scale commercial operations. [23]

In Thailand I enjoyed eating fish, chicken and sea food, pork, rice and lots of vegetables and fruits. Beef was really not on the menu. A story of why beef was not my favorite food then. Nearing the end of a trip I visited a steak house near my residence. This particular evening, I decided to visit this steak house and have a steak. On the menu was an Australian steak, but it tasted like shoe leather. This steak was so tough, I could not cut into the meat. I got up, paid my bill and never went back to the restaurant.

Upon returning back to the United States, my wife and decided to go to a local French restaurant and have filet mignon steak with a delightful pepper sauce. The steak was absolutely scrumptious, you could cut it with your fork. When I went home though,

---

21 https://www.tandfonline.com/doi/abs/10.1080/09700161.2022.2111765?journalCode=rsan20
22 https://courses.lumenlearning.com/suny-environmentalbiology/chapter/9-5-sustainable-agriculture/
23 https://wonderfruit.co/wonderpost/supporting-farming-communities-during-a-crisis/

my stomach couldn't digest the steak because I had not eaten red meat for so long my digestive system couldn't handle it. At this time, I decided to stop eating beef. Don't get me wrong, I still love a good hamburger once in a while.

We planned a trip to Greece for a holiday 2009. Arriving in Athens, found a small restaurant that had stayed open late. They had cheeses and other assorted deli products. We ordered food and enjoyed our evening snack. On this first encounter of eating food outside the United States and Thailand, it tasted better and looking back, I had no allergic reactions!

This new gastronomical experience was extremely pleasant. We did some island hopping to Santorini, Paros, Mykonos, and finally to Corfu. The entire 3 ½ weeks in Greece I was able to eat fresh vegetables, meats, and seafood without an allergic reaction. Returning to the United States, the rashes would soon reappear. Chalking the allergy up to coincidence at the time and not putting anything together about food from another country verses food here in the United States.

In 2012 my wife and I returned to Thailand for a month-long holiday in Phuket, for business and pleasure. We set our plans to go during Christmas and New Year's holiday. We stayed at a place called the Sawadee hotel in Phuket. This was one of the best holidays we had outside the country in years. The ambiance of this hotel is very unique, suites with a private pool and other wonderful amenities.

Each of the two-story suites had one enormous king size bed with a gold dome embedded into the ceiling over the bed, and a water foot massager at the end of the bed. This oversized king size bed could have slept at least four people. Ornate, sculptured Sliding doors separate the bedroom from the bathroom. Downstairs had a kitchen and another shower to go into the pool.

In the evening, we would go to local restaurants in Phuket to enjoy the atmosphere of the area. None of these foods caused any skin reactions. My rosacea that shows up in the United States disappeared again. The entire trip I was able to eat everything without a food allergy.

I am not alone! I heard stories about experiences similar; travelers to Europe whose food allergies went away when they were outside the US.

After multiple trips to Thailand, I realized my rosacea or was cleared up on every trip. Each trip returning to the United States rosacea returned to my face, as I've been dealing with for over 20 years. I have poured more chemicals on my face prescribed by dermatologists than I care to think about. Thousands and thousands of dollars were spent on creams and lotions to help clear up so-called acne or rosacea.

In 2019-2020 we planned another long trip, after more than 3 years fighting my rash. We wanted to see if eating food in Thailand again would again have a good effect on my body.

Good news! I started my food experiment at the Boathouse hotel buffet. This buffet spread would delight almost anyone. I chose the buffet and a homemade omelet. On the bread table was a large assortment of breads and pastry all made at the hotel. My choice was 2 fresh croissants and fresh cooked crispy bacon. My wife said to me: you're going to be sorry for eating all that, you are going to get allergic reaction. I said we'll have to wait and see . We finished our delicious breakfast and went for a walk on the beach for about an hour. Upon returning back to the hotel, I had no allergic reaction from eating the food. My experiment on myself worked: the bacon and the bread did not cause an allergic reaction.

This is an expensive trip to find out if you're allergic to food in a foreign country. For the rest of the trip, I had no food allergic reactions or rashes on my legs. My food experiment was working. Basically, by the end of our 30 days in Thailand, I was able to eat anything without any rashes or itching. My wife was impressed, and I was delighted. I was impressed as my food experiment worked 8,000 miles from home. Now facing the end of our trip and going back home to uncertainty of eating food with pesticides. This was an experiment in Thailand, a beautiful country to visit but not my home. Facing reality of going back to my limited diet In the US and not being able to eat so many different kinds of food is a dreadful truth.

About a month after our return from Thailand along came Covid. My diet went back to eating salmon and chicken and the recipes I have shared (see appendix). Going shopping in a quasi-lockdown was a tremendous burden to each and every one of us. While shopping, no one interacted, we all seemed to be zombies in our own little cocoon space. While wearing masks, the one thing I missed the most was not seeing people smile out in public.

We went back to Thailand in September - October 2022 for five weeks. Again, the purpose of this trip was to try the food in Thailand to see if I would have any food reactions. The trip was a success; again, as I could eat everything. A bonus was my blood pressure dropped to 111 / 70 and my doctor tells me this is wonderful. And I did not have to take any blood pressure medication during this entire trip.

In 2017 evidence for a correlation of high blood pressure and pesticides was published in the Journal of Occupational and Environmental Medicine. [24] The study found that individuals who were exposed to certain classes of pesticides, including organophosphates and pyrethroids, had a higher risk of developing hypertension. In 2018, the Journal of the American Society of Hypertension found a link between pesticide exposure and increased blood pressure levels in agricultural workers. [25]

In 2020, Thailand did ban the use of three pesticides - Chlorpyrifos, Paraquat, and Glyphosate - due to health and environmental concerns. These pesticides are known to be hazardous to human health and have been linked to various health problems such as cancer, birth defects, and neurological disorders. Thailand's decision to ban these pesticides is in line with a global trend to reduce the use of harmful chemicals in agriculture, by promoting sustainable farming practices.

---

24 https://journals.lww.com/joem/Abstract/2017/11000/Occupational_Exposures_and_Metabolic_Syndrome.4.aspx
25 Journal of the American Society of Hypertension 2018

In summary, when I visited countries outside of the United States where pesticides are banned, (Europe has banned 464, and some countries in Asia have many bans on pesticides as well) I had zero rashes occur on my body. Yes, this is not a scientific study, this is a study of one person's experience with food, organic and non-organic.

Another reason pesticides are not used in countries like Thailand is they are expensive to apply. This expense is actually a good thing as small farmers in foreign countries cannot afford to apply these chemicals.

Another beneficial experience I had when outside the US: my blood pressure dropped under 120/70, pretty good BP right? Back in the United States, if I do not eat organic food, my blood pressure can go to a high of 180/85, getting dangerous. After a visit to my doctor, I had to limit my consumption of red wine, green tea, etc. I also take blood pressure (BP) medication losartan 25mg twice daily.

I know my own body and pay attention to my health. My allergic reaction to pesticides has forced me to pay more attention to my health. Of course, each of us has different body types, and each person needs to consult their doctor to make an intelligent decision.

Finding organic foods in the US is a challenge. Over the last few years, there has been greater attention to people like me who are asking for organic food to be available in the grocery store. However, organic supply is still limited, and the grocery stores raise prices on organic foods, sometimes more than double of non-organic.

Pollinators

BEES POLINATE

World Bee Day is celebrated on May 20th each year. Put this in your calendar. [26]

Our world's honeybee population is deteriorating at a rate never seen before. Honeybees are crucial to the pollination of crops. Giant farms plant one crop for the highest yield. New studies now appearing in journals show crop diversification can eliminate the use of pesticides. For example, Sharing the farmland with animals that can eat the bugs and provide a protein source at the same time. We can make farming more efficient, and society needs the will to try.

We all enjoy nature: seeing flowers or having a cup of coffee, tea, a hand full of nuts, vine ripe tomato, or a juicy apple. All are pollinated by bees. Bees complete this transfer of pollen from one flower to the next.

According to the United Nations Environment Program, of the 100 crop varieties there are over a hundred different crop varieties providing 90% of the world's food supply. Seventy-one percent of these crops are pollinated by bees. [27] When you eat your next vegetables or fruits (squash, tomatoes, cherries, blue-berries, almonds, avocados, cranberries, or apples) thank bees for their pollination efforts. In North America, honeybees alone pollinate nearly 95 different types of fruits, in addition to crops like soy and corn. [28]

Three out of four crops across the world, delivering vegeta-bles or fruits or seeds for human consumption depend, at least in part, on pollinators and bees do a lot of this work. [29] In our small

---

26 https://www.fao.org/3/i9527en/i9527en.pdf
27 https://www.fao.org/3/i9527en/i9527en.pdf
28 https://www.centerforfoodsafety.org/issues/304/pollinator-protection/impacts-on-the-food-supply
29 https://ourworldindata.org/pollinator-dependence

world there are over 20,000 different species of bees pollinating flowers and crops and 4,000 of these bees are native to north America. [30] Some native bees are so specialized they restrict pollination to certain plant species. [31] Bees come in a variety of sizes, colors, and shapes because there are a variety of flowers and colors to be pollinated. Science is still researching native bees as there is a lot to learn. It is estimated 10% of native bees in north America are still unidentified as pollinators. [32]

Bees increase crop yields by about 24 per cent and increase crop value by $15 billion each year in the US alone. Bees are essential to our food system with 87 of the world's leading crops requiring pollination, making bees' contribution to agriculture a necessity for global food security. [33]

Bees get help with pollination from some moths, flies, wasps, beetles, ants, and butterflies as well as some animals pollinating plants. Vertebrate pollinators include bats, non-flying mammals, including several species of monkeys, rodents, lemur, tree squirrels, chipmunks, honey possums, Olingo (central America) and kinkajou (central and parts of south America). In addition, there are there are about 2,000 species of pollinating birds as hummingbirds, sunbirds, honeycreepers and parrot species like lories and lorikeets native to Australia. [34] One study found more than 25 different bees and hoverfly species on apple blossom and showed that insect pollination tripled the production of fruit and boosted its size and quality. Nature can do a better job without pesticides.

For millions of years, bees, other insects and animals have helped nature and humans survive. Civilizations have grown corn and soybeans along with other fruits and vegetables to allow mankind to achieve what has been accomplished. As an agrarian

---

30 https://www.usgs.gov/faqs/how-many-species-native-bees-are-united-states
31 https://www.usgs.gov/faqs/how-many-species-native-bees-are-united-states
32 https://www.usgs.gov/faqs/how-many-species-native-bees-are-united-states?items_per_page=6&page=1
33 https://obamawhitehouse.archives.gov/the-press-office/2014/06/20/fact-sheet-economic-challenge-posed-declining-pollinator-populations
34 https://www.fs.usda.gov/wildflowers/pollinators/animals/birds.shtml

society, tilling the soil moving forward has been able to grow hydroponics food in an enclosure and not need pesticides. However, today our world is now experiencing unintended consequences from one pesticide glyphosate that has been used for 50 years.

Remember our basic biology about pollination is transferring pollen grains from the male anther to a flower with female stigma. Honeybees complete this transfer of pollination from one flower to the next. Bees increase crop yields by about 24 per cent and increase crop value by $15 billion each year in the US alone. Bees are essential to our worlds food system by providing 90% of food pollination as a necessity for global food security.

Or another way to look at pollination is three out of four crops across the world one of 20,000 different species of bees are performing pollination and 4,000 are native to north America. [35] Some native bees are so specialized they restrict pollination to certain plant species. [36] Many of our native wild and crop plants have sets of bees that are so specialized that they restrict their visits to those plants alone.

News flash: one of 3 bites of food we eat comes from crops pollinated by honeybees. Over the past decade honeybee population have decreased in the United States by 44 percent. From 2015 through 2016 some beekeepers announced they lost all of their honeybee populations.

The type of pesticides used on honeybees causing their decline are nicotine-based systemic insecticides called neonicotinoids. Neonicotinoids are agricultural insecticides resembling nicotine and are one of the most used insecticides in the world. When honeybees become exposed to neonicotinoids through nectar, pollen, on plants it interferes with their nervous system causing tremors, paralysis, and death. One treatment of neonicotinoids on corn contains enough insecticide to kill 80,000 honeybees. [37]

---

35 https://www.usgs.gov/faqs/how-many-species-native-bees-are-united-states
36 https://www.usgs.gov/faqs/how-many-species-native-bees-are-united-states
37 https://www.centerforfoodsafety.org/fact-sheets/2093/pollinators-and-pesticides

The average frame size of a honeybee hive count bees 19 inches wide and nearly 8.5 inches deep, there are enough cells to produce 3,500 or more bees per side of comb, or about 7,000 adult bees per frame. Beehives contain 8 to 10 of these frames hence there is an educated guess as to the number of honeybees in each hive. Data here shows why one spray on corn of neonicotinoids can kill an entire honeybee hive. [38]

BIRDS POLLINATE

Birds are incredibly important pollinators of wildflowers, playing a critical role in many ecosystems throughout the world. In North America alone, there are over 2,000 different species of birds that are known to pollinate flowers. These birds not only provide essential pollination services but also contribute to the overall health and diversity of our ecosystems.

Many different types of birds are known to pollinate flowers, but some of the most common include hummingbirds, honeyeaters, spiderhunters, sunbirds, and honeycreepers. These birds are often attracted to flowers with brightly colored petals and a sweet nectar that provides them with the energy they need to fly long distances.

The process of birds pollinating flowers is called ornithophily, and it plays a vital role in the pollination of many plant species. While some flowers are pollinated by insects, such as bees and butterflies, many others rely on birds to transfer their pollen from one flower to another.

Nature has developed a clever strategy to attract birds to flowers. Many flowers are brightly colored and have a unique shape that makes them easy for birds to identify. They often produce nectar, a sweet liquid that provides birds with a high-energy food source. By consuming the nectar, birds inadvertently pick up pollen from the flower and transfer it to other flowers, allowing for pollination to occur.

---

38 https://galenafarms.com/blogs/articles/why-choose-an-8-frame-beehive

Unfortunately, the use of pesticides can have a devastating effect on the habitats that support birds that pollinate wildflowers. Pesticides such as herbicides can destroy nearby vegetation that provides food and shelter for birds. This, in turn, can lead to a decline in the number of birds that are available to pollinate wildflowers, which can have a significant impact on the local ecosystem.

This is an unintended consequence of using pesticides to remove unwanted vegetation. While the primary goal of using herbicides may be to remove weeds or unwanted plant species, the unintended consequence of this action can be the depletion of wildlife vegetation, which eventually depletes the bird's pollination of wildflowers.

To protect the habitats that support bird pollination, it is important to use alternative methods for managing vegetation, such as hand-weeding or using organic herbicides. By doing so, we can help to preserve the delicate balance of our local ecosystems and ensure the continued pollination of wildflowers by birds.

In addition to pollination, birds also play an important role in seed dispersal. After consuming the nectar of a flower, birds may also eat the fruit or seeds of the plant. As they fly from one location to another, they may inadvertently drop the seeds, helping to spread the plant throughout the ecosystem.

Birds are also important indicators of ecosystem health. Changes in bird populations can be an early warning sign of problems within the ecosystem, such as habitat loss, pollution, or climate change. By monitoring bird populations, scientists can gain insights into the overall health of the ecosystem and take steps to address any issues that may arise.

Overall, birds are critical pollinators of wildflowers, and they play a vital role in many ecosystems throughout the world. By protecting their habitats and reducing the use of harmful pesticides, we can help to ensure the continued health and diversity of our ecosystems and the important role that birds play within them.

Hummingbirds are some of the most fascinating and important pollinators in the world. With more than 300 different species of hummingbirds found across the Americas, from Alaska to Chile, they are found in a wide variety of habitats, including forests, deserts, and even urban areas. [39]

Hummingbirds play a crucial role in wildflower pollination, visiting flowers to feed on nectar and inadvertently transferring pollen between plants. This pollination is essential for the plants to produce fruit or seeds and maintain healthy populations. In fact, some plants have evolved specifically to attract hummingbirds, with brightly colored flowers and long, tubular shapes that are perfectly adapted for the birds' long, thin bills.

Despite their small size - weighing in at less than a penny - hummingbirds have a voracious appetite, consuming up to twice their body weight in nectar each day. In addition to nectar, hummingbirds also eat insects like ants and gnats, which provide essential protein for their growth and development.

Unfortunately, the use of pesticides on plants can disrupt the diet of hummingbirds, either by killing off the insects that they rely on or by contaminating the nectar that they feed on. This can have serious consequences for the health and well-being of hummingbirds and can also disrupt the delicate balance of the ecosystem as a whole.

As such, it is important to be mindful of the impact that pesticides can have on hummingbirds and other important pollinators. By using organic gardening methods and avoiding harmful pesticides, we can help to ensure that hummingbirds and other pollinators have access to the food and habitat they need to thrive.

Hummingbirds are not the only important pollinators in the world. In fact, there are many different species of birds that play a crucial role in pollination in various regions of the planet. For example, in Hawaii, the native honeycreepers are an important

---

39 https://sciencing.com/hummingbirds-help-pollination-4566837.html

group of bird pollinators, with their long, curved bills adapted for feeding on nectar from flowers. Similarly, in Australia, honeyeaters are important pollinators, feeding on nectar and transferring pollen between plants as they go.

In New Guinea, brush-tongued parrots are important pollinators of a variety of plant species, while in Southeast Asia, sunbirds play a similar role, visiting flowers to feed on nectar and transferring pollen between plants in the process. In Southern Africa, a variety of birds, including sunbirds, sugarbirds, and white-eyes, are important pollinators of a wide range of flowering plants.

In Japan and India, a variety of bird species are important pollinators, with bulbuls, flowerpeckers, and sunbirds among the most commonly seen at flowering plants. These birds visit flowers to feed on nectar and pollen, and in the process, help to ensure that the plants are able to reproduce and maintain healthy populations. [40]

Overall, birds play a critical role in pollination around the world, with a wide variety of species adapted to different habitats and ecosystems. By understanding and protecting these important pollinators, we can help to ensure that the plants we rely on for food and other resources are able to thrive and support healthy ecosystems for generations to come.

In conclusion, ornithophily plays a vital role in the pollination of many plant species, and it is essential that we protect the habitats that support this process. The unintended consequences of using pesticides to remove unwanted vegetation can have a significant impact on the ecosystem, leading to a decline in the number of birds available to pollinate wildflowers. By using alternative methods for managing vegetation, we can help to preserve the delicate balance of our local ecosystems and protect the important role that birds play in pollinating wildflowers.

____

[40] https://www.fs.usda.gov/wildflowers/pollinators/animals/birds.shtml

Pesticides and bird population is not a good combination. From 2008-2014 on a county recording keeping study and observed bird route conducted by the U.S. Geological Survey (USGS). The USGS made 15,000 observations that statically analyzed bird-pesticide pairs in the centennial 48 states. This study was to understand neonicotinoid use and decline in bird biodiversity. These results showed that only a one-hundred-kilogram increase in neonicotinoid use in one county with grass land birds decreased the bird's population by 2.2 percent and non- grass-land birds by 1.4 percent. [41] The USGS made 15,000 observations that statically analyzed bird-pesticide pairs in the contiguous 48 states. This study was to understand neonicotinoid use and decline in bird biodiversity. The USGS study also found that the effects of neonicotinoid pesticides on birds varied depending on the type of bird and the habitat in which it lived.[42] For example, grassland birds such as meadowlarks and sparrows were more negatively impacted by neonicotinoids compared to non-grassland birds like blue jays and robins. This is because grassland birds rely on insects that are more likely to come into contact with neonicotinoid-treated seeds or soil, while non-grassland birds tend to feed on a wider variety of foods. [43] Additionally, the study found that neonicotinoids had a greater impact on bird populations during the breeding season, which is when birds need to find and consume large amounts of insects to feed their young. Overall, the study underscores the importance of considering the unintended consequences of pesticide use on wildlife and the environment. [44]

This decline in bird population could have significant ecological and economic impacts. For example, grassland birds play an important role in controlling insect populations and promoting

---

41 https://www.sciencedaily.com/releases/2020/08/200814131023.htm
42 https://ocm.auburn.edu/newsroom/news_articles/2020/10/141359-miao-bird-study.php
43 https://www.sciencedaily.com/releases/2020/08/200814131023.htm
44 https://www.smithsonianmag.com/science-nature/popular-pesticides-linked-drops-bird-population-180951971/

plant growth by consuming insects that can damage crops and other vegetation. Without these birds, there may be an increase in insect populations and a decrease in crop yields. This could in turn affect the availability of resources such as timber, clean water, and wildlife habitat, which are important for human health and well-being. Therefore, it is crucial to consider the potential long-term consequences of pesticide use on bird populations and take steps to mitigate the impact on the environment.[45]

Basically, these scientists were studying a select number of bird species and we need to ask what happens to the other species where neonicotinoids are applied to fields. Here is what the EPA has confirmed Three Widely Used Neonicotinoid Pesticides Likely Harm Vast Majority of Endangered Plants, Animals. [46]

The impact of these pesticides on endangered species and their habitats is a major cause for concern. The statistics presented show that the use of imidacloprid, thiamethoxam, and clothianidin can have devastating effects on a large number of plants and animals. The fact that nearly 80% of all endangered species are likely to be "adversely affected" by imidacloprid alone is alarming. This means that the use of these pesticides can contribute to the extinction of species and disrupt the balance of ecosystems. Moreover, the modification of critical habitats for these species can further reduce their chances of survival. It is important to consider the potential long-term consequences of using these pesticides and to explore alternative methods of pest control that are less harmful to the environment and the species that inhabit it. [47]

Neonicotinoid pesticides have been found to have harmful effects on bird populations. In addition to anorexia, these pesticides have been linked to decreased bird population numbers and

45 https://ocm.auburn.edu/newsroom/news_articles/2020/10/141359-miao-bird-study.php#:~:text=When%20birds%20eat%20the%20pesticide,decrease%20birds'%20abilities%20to%20reproduce
46 https://biologicaldiversity.org/w/news/press-releases/epa-confirms-three-widely-used-neonicotinoid-pesticides-likely-harm-vast-majority-of-endangered-plants-animals-2022-06-16/
47 https://biologicaldiversity.org/w/news/press-releases/epa-confirms-three-widely-used-neonicotinoid-pesticides-likely-harm-vast-majority-of-endangered-plants-animals-2022-06-16/

behavioral changes, including changes in vocalizations, feeding habits, and nesting behavior. One study found that neonicotinoids also affected the orientation of migratory birds, causing them to become disoriented and lose their way. This can lead to birds becoming lost and disoriented, which can impact their ability to complete their migrations and find suitable breeding and nesting grounds.

The European Union banned the use of neonicotinoids in 2018 due to concerns about their impact on pollinators and other wildlife. The United States has been slower to act, but some cities and states have implemented restrictions on neonicotinoid use. However, there are still over 1,000 different pesticides in use today, and some companies are able to mix multiple pesticides together in a single vat. This means that simply banning neonicotinoids may not be enough to protect bird populations and other wildlife from the harmful effects of pesticides. Further research and regulation is needed to ensure that all pesticides are safe for use and do not harm the environment or wildlife. [48]

The unintended consequences of using synthetic pesticides can have severe implications for our environment and health. When pesticides are used, they not only kill targeted pests but can also harm non-targeted species, such as beneficial insects like bees and butterflies, birds, and aquatic organisms. Additionally, these chemicals can seep into the soil and water, contaminating our food, drinking water, and the environment. Long-term exposure to pesticides has been linked to various health problems, such as cancer, birth defects, and neurological disorders.

The impact of pesticides on bird populations is a major concern for conservationists and ecologists. The decline of predatory birds like the Brown Pelican, Osprey, Bald Eagles, Peregrine Falcons, Swinson hawks, and many other bird species is directly linked to the widespread use of pesticides. These birds, along with

---

48 https://www.science.org/content/article/common-pesticide-makes-migrating-birds-anorexic

other raptors, are at the top of the food chain and are highly susceptible to pesticides because they consume prey that has been contaminated with these chemicals. The loss of these birds has had a significant impact on the ecosystems they inhabit and has disrupted the natural balance of many food webs. [49]

Moreover, the negative impact of pesticides on bird populations is not limited to predatory species. Studies have shown that pesticide use on agricultural lands has led to a ten percent decline in bird populations in the United States. This is because many bird species rely on insects and other small creatures that are killed by pesticides, leading to a loss of food and habitat. In addition, many bird species migrate long distances and can be exposed to pesticides in different parts of the world, leading to a decline in populations across large regions.

By reducing our reliance on pesticides and adopting more sustainable pest management practices, we can help protect bird populations and maintain healthy ecosystems. Diversifying crops and practicing crop rotation are effective methods to reduce pesticide use and improve soil health. When a farmer grows a variety of crops instead of just one, they can break up pest and disease cycles, reduce the need for pesticides, and improve soil fertility. Crop rotation, which involves changing the crops grown in a particular field over time, also helps break up pest and disease cycles and can improve soil health by adding different nutrients and organic matter.

However, for farmers to implement these practices on a large scale, there needs to be government support in the form of subsidies and crop insurance. Currently, the government provides subsidies for farmers who grow certain crops, which often leads to a monoculture approach to farming and increased pesticide use. By incentivizing farmers to diversify their crops and reduce pesticide use, the government can help protect the environment and promote sustainable agriculture.

---

49 https://nationalzoo.si.edu/migratory-birds/news/when-it-comes-pesticides-birds-are-sitting-ducks

Overall, reducing pesticide use is a complex issue that requires a multifaceted approach, including government support, farmer education, and innovative pest management strategies.

In addition to providing habitat for multiple species, diversifying crops can also improve soil health and reduce the need for pesticides. Crop rotation, a technique that involves changing the type of crop grown in a particular field each year, can help reduce soil-borne diseases and pests, as well as improve soil fertility. Cover crops, which are grown between planting seasons to help protect and enrich the soil, can also be used to suppress weeds and pests.

Furthermore, diversifying crops can have economic benefits for farmers. By cultivating a diverse range of crops, farmers can lessen their reliance on a single commodity and broaden their profit potential by marketing a wider array of products. This can also help to build more resilient and sustainable farming communities. [50]

However, diversifying crops and implementing sustainable farming practices requires investment and support from policymakers, researchers, and the agricultural industry. This may include funding for research and development of sustainable farming techniques, education and training for farmers, and policies that incentivize sustainable practices and support small and medium-sized farmers.

INSECTS POLLINATE

In addition to the pollinators mentioned above, there are also some surprising pollinators that most people are not aware of. For example, moths, flies, wasps, beetles, ants, and butterflies as well as some animals pollinate plants some reptiles and amphibians can also help with pollination, such as lizards and tree frogs. Certain species of fish and even some mammals that feed on nectar, such as marsupials, can also contribute to pollination.

---

50 https://www.ncbi.nlm.nih.gov/pmc/articles/PMC7078872/

In some regions, such as in Madagascar, lemurs have co-evolved with certain plants to become their primary pollinators, making them essential for the survival of those plant species. The diversity of pollinators is critical for maintaining the health and resilience of ecosystems around the world, and efforts should be made to protect and support these important species. Vertebrate pollinators include bats, non-flying mammals, including several species of monkey, rodents, lemur, tree squirrels, chipmunks, honey possums, Olingo (central America) and kinkajou (central and parts of south America). In addition, there are there are about 2,000 species of pollinating birds as hummingbirds, sunbirds, honeycreepers and parrot species like lories and lorikeets native to Australia. [51]

While insects travel from one flower to the next their body and legs transfers the pollen that is sticking to them to male parts to the female part of the flower thus pollination occurs.[52] One study found more than 25 different bees and hoverfly species on apple blossom and showed that insect pollination tripled the production of fruit and boosted its size and quality. Nature can do a better job without pesticides.

There are 12 states that will not protect insect pollinators: Alabama, Alaska, Arizona, Colorado, Indiana, Nevada, North Carolina, Oregon, Pennsylvania, Utah, West Virginia, and Wyoming. [53] Their laws do not allow insect pollinators to be protected as a species. Without pollinators humans as a species will vanish from the planet. There are an estimated 1,400 crops planted on earth to feed humans and 80% of these plants require pollination by insects and animals. In the United States 150 food crops depend on pollinators, including almost all fruit and grain crops.

These native pollinators are worth an estimated $3 billion per year in the United States. Pollinators are free and the those growing crops are benefiting from all this free pollination labor. [54]

---

51 https://www.fs.usda.gov/wildflowers/pollinators/animals/birds.shtml
52 https://www.buzzaboutbees.net/insect-pollination.html
53 https://nyti.ms/3Jfn5f7
54 https://www.fs.usda.gov/wildflowers/pollinators/importance.shtml

If you want to help these insects, grow pollinator friendly wildflower plants in your home or garden. In Europe, 84% of their crops and the vast majority of wildflowers depend on insect pollination. In the United States, farmers are applying vast amounts of pesticides to remove insects that could help increase crop production.[55]

Insects matter in biodiversity and make up 80% of species on earth. They play a critical role in supporting life as we know it in our world. Insects are decomposers of organic material and help rejuvenate the soil where our crops are grown. Multiple species depend on insects for their existence: reptiles, amphibians, some bird species, along with hedgehogs, moles, shrews, anteaters, armadillos, and bats.

MONARCH BUTTERFLIES POLLINATE

Research on pesticides helps us understand the scale and impact of the problem. When reading various articles about pesticides, it becomes clear that many species are affected by their use. Pesticides can harm bees, monarch butterflies, birds, insects, mammals, and amphibians, among others. It's important to recognize the broader impact of pesticides beyond just their intended target.

The monarch butterfly is one of the most recognizable and extensively studied butterfly species in North America. Many conservationists, government agencies, schools, researchers, and ordinary people track the breeding, migration, and wintering habitats of these butterflies. [56]

The monarch butterfly has been on a decline for decades in North America. Multiple reasons have contributed to this loss of the monarch butterfly ecosystems, commercial agricultural practices and even conventional gardening contribute to the decline of this iconic species. To get involved got to the National Wildlife Federation and see their comprehensive campaign to help save

---

55 https://www.nhm.ac.uk/discover/insect-pollination.html
56 https://www.fs.usda.gov/wildflowers/pollinators/Monarch_Butterfly/

the Monarch Butterfly.[57] In addition, there are other programs to help save the monarch butterflies working with between Journey North, a program of the University of Wisconsin-Madison Arboretum, and Monarchs Across Georgia, and The Environmental Education Alliance of Georgia to mention a few. [58]

The monarch butterfly is a well-known and beloved species that has captured the hearts and imaginations of people all over the world. Unfortunately, these beautiful creatures are facing multiple threats to their survival, including the widespread use of pesticides and the destruction of their habitat due to increased agriculture and development.

One of the most critical issues facing the monarch butterfly today is the loss of milkweed plants, which are essential for their reproduction and survival. Milkweed is the only plant that monarch butterflies can lay their eggs on, and the loss of these plants is having a devastating impact on the monarch population. In North America, where monarch butterflies are most commonly found, the use of pesticides has resulted in the destruction of many milkweed plants, which has led to a decline in the number of monarch butterflies in the wild.

In addition to the loss of milkweed plants due to pesticide use, the destruction of natural habitats is another major threat to the monarch butterfly population. With an increasing demand for farmland, many natural habitats are being tilled to make room for crops, leaving the monarch butterfly with fewer places to lay their eggs and fewer milkweed plants to sustain their offspring.

As a result, monarch butterflies are becoming an endangered species, and urgent action is needed to help protect them.

The milkweed plant genus Asclepias is a perennial flowering plant native to North America, and there are twelve species of milkweed plants that monarch butterflies use to lay their eggs. [59] A single female monarch butterfly can lay up to 500 eggs on a single

---

57 https://blog.nwf.org/2015/02/twelve-native-milkweeds-for-monarchs/
58 https://www.thebutterflynetwork.org/program/journey-north
59 https://en.wikipedia.org/wiki/Asclepias

milkweed plant, which demonstrates just how critical these plants are to the survival of the species. Without access to milkweed plants, the monarch butterfly population will continue to decline, and the future of this beautiful and beloved species is at risk.

The loss of milkweed plants and destruction of natural habitats are significant threats to the survival of the monarch butterfly. It is crucial that we take action to protect these essential plants and preserve the habitats that monarch butterflies need to thrive. By working together, we can help to ensure the survival of the monarch butterfly and ensure that future generations can enjoy the beauty of this remarkable species. [60]

The journey of the monarch butterfly is nothing short of remarkable. Every year, millions of monarch butterflies embark on a migration journey that takes them from North America to their wintering grounds in Mexico. This journey can cover over 3,000 miles and can take up to 45 days to complete.

Once the monarch butterflies arrive at their wintering grounds in the oyamel fir forests of southwestern Mexico, they congregate in enormous clusters, covering the trees and turning the forest into a breathtaking sea of orange and black. Of the 12 winter mountain sites that monarchs use, four are dedicated preserves, located approximately two hours outside Mexico City. These sanctuaries are protected areas where visitors can see the butterflies up close and learn about their remarkable journey and the importance of protecting their habitat. [61]

At these sanctuaries, a docent will accompany visitors to help protect the monarch butterflies and the forest for future generations. [62] The docent will provide valuable information about the monarch butterflies' behavior and the ecosystem in which they live, helping visitors to understand the importance of preserving these habitats.

---

60 https://blog.nwf.org/2015/02/twelve-native-milkweeds-for-monarchs/
61 https://journeynorth.org/tm/monarch/SanctuaryFactsOyamel.html
62 https://middlejourney.com/seeing-the-monarch-butterfly-migration-in-mexico/

While the oyamel fir forests of southwestern Mexico are the most well-known wintering grounds for monarch butterflies, there are other migration hotspots in North America. Texas and California are two other popular destinations for monarch butterflies during their migration journey, and some monarchs have even reached overwintering sites in California. [63]

In conclusion, the migration journey of the monarch butterfly is a fascinating and awe-inspiring spectacle that demonstrates the resilience and adaptability of this remarkable species. By protect ing their habitats and preserving their migration routes, we can help to ensure the survival of the monarch butterfly for genera-tions to come.

The Monarch Milkweed Project is an initiative designed to help protect and support the monarch butterfly population by providing guidance on what types of milkweeds to plant in your garden or patio. As mentioned earlier, milkweed is the only plant that monarch butterflies can lay their eggs on, and without access to this plant, the species is at risk of extinction.

The project aims to educate people about the importance of milkweed and how to create a monarch-friendly habitat. By plant-ing milkweed in your garden or patio, you can help support the monarch butterfly population and provide a safe haven for these remarkable creatures.

The rapid decline of the monarch butterfly population has made the Monarch Milkweed Project more important than ever before. The United States Department of Agriculture (USDA) has described the monarch butterfly as the most spectacular insect in the world and an endangered species. [64] The monarch population has declined by more than 90 percent in recent years, and urgent action is needed to help protect these beautiful creatures.

---

63 https://journeynorth.org/monarchs/news/fall-2022/10102022-magical-migration
64 https://www.fs.usda.gov/wildflowers/pollinators/Monarch_Butterfly/

The Monarch Milkweed Project is a vital initiative that can help to protect and support the monarch butterfly population. [65] By participating in the Monarch Milkweed Project, you can help to make a difference in the survival of the monarch butterfly. The project provides valuable resources and information on what types of milkweeds to plant, how to care for your milkweed plants, and how to create a monarch-friendly habitat in your yard. Start planting in the fall to get your perennial milkweed varieties started. The milkweed seeds need a cold season to get started. Most of the time the winter season will take care of this naturally. Find out what kind of milkweed plants are native to your region of the country for the best results. [66] [67] The milkweed project identifies twelve native milkweeds for Monarch butterflies that can help them survive. [68] (See the appendix "Milkweed you can plant by region" to find out what types of milkweed plants grow in your region)

Since we all enjoy seeing monarch butterflies, speak to your local farmers and encourage them to set aside acreage to plant milkweed for the monarch butterflies to be revitalized. Grow milkweed in your backyard or a small patio. Help save the important-monarch butterfly!

We should be concerned as a society that mother nature will win and all the creative scientists can do is try and outsmart mother nature but in the end mother nature will win.

It's important to recognize the broader impact of pesticides beyond just their intended target.

---

65 https://monarchmilkweedproject.org/
66 https://monarchbutterflygarden.net/fall-planting-milkweed-10-steps/
67 https://www.nps.gov/subjects/pollinators/migratingmonarchs.htm
68 https://blog.nwf.org/2015/02/twelve-native-milkweeds-for-monarchs/

Toxic effects of pesticides

The use of DuoEnlist™ technology has sparked concerns about the potential negative impacts on human health and the environment. The increased use of 2,4-D, glyphosate, and fop herbicides in combination with genetically engineered seeds has led to the emergence of new superweeds that are resistant to multiple herbicides. The use of multiple herbicides also increases the risk of herbicide drift and contamination of nearby crops, water sources, and ecosystems. In addition, 2,4-D has been linked to various health problems, including cancer, reproductive and developmental issues, and immune system dysfunction. Despite these concerns, the pesticide industry continues to promote the use of DuoEnlist™ technology as a solution to weed resistance, without fully understanding the long-term consequences of this approach on human health and the environment. As consumers, it is essential to be aware of the potential risks associated with the use of pesticides and advocate for more sustainable and environmentally friendly agricultural practices. [69]

Again, we have a pesticide banned in the European Union and here in the United States it is still the wild west pesticide show and which company can be the best gun slinger selling and promoting pesticides.

Toxicity Human exposure to 2,4-D has substantially risen despite a multitude of health and environmental concerns. Another fact 2,4-D applied in agriculture increased 67% between 2012 and 2020. [70]

Herbicides have been a widely used tool for controlling vegetation in various settings, including residential, commercial, and industrial. They are applied to control weeds, invasive species, and other unwanted vegetation that can affect the aesthetics of

---

69 https://ehjournal.biomedcentral.com/articles/10.1186/s12940-021-00815-x
70 https://www.theguardian.com/environment/2022/feb/09/toxic-herbicide-exposure-study-2-4-d

landscapes and disrupt ecosystem functions. However, the use of herbicides also poses significant risks to the environment, particularly to aquatic systems.

When herbicides are applied to water bodies or near aquatic habitats, they can cause harm to fish, amphibians, and other aquatic organisms, which can have long-lasting impacts on the health of ecosystems. Herbicides can alter the water chemistry and reduce the availability of oxygen for aquatic organisms, leading to death or reduced growth rates. They can also bioaccumulate in aquatic organisms, which can ultimately impact human health when these organisms are consumed.

Additionally, herbicides can persist in the environment for extended periods, contaminating soil and water and posing risks to human health. Long-term exposure to herbicides has been linked to various health problems, including cancer, reproductive issues, and developmental delays.

Given the potential risks associated with herbicide use, it is essential to consider alternative methods of controlling unwanted vegetation. These methods include the use of mechanical controls, such as mowing and trimming, as well as biological controls, such as the introduction of natural predators or competitors of invasive species. By implementing such alternatives, we can reduce the potential environmental impacts of herbicides and promote sustainable management of landscapes and ecosystems.[71] An obvious question should they be applied at all.

There are six systemic categories to Toxicity levels: *Acute, Subchronic, Chronic, Carcinogenicity, Developmental, and Genetic (somatic cells).*

1. Acute toxicity occurs almost immediately within (seconds/minutes/hours/days) after an exposure. An acute exposure is usually a single dose, or a series of doses received within a 24-hour period. Death can be a major concern in cases of acute exposures. One pesticide, methyl isocyanate, is used in the production

---

71 https://www.epa.gov/caddis-vol2/herbicides

of pesticides. From an industrial accident in India in 1989 methyl isocyanate killed 5,000 people and an additional 30,000 plus were permanently disabled.

2. The second most dangerous toxic exposure is called Subchronic pesticides and occurs form repeated exposure over several weeks or months. Some categories are organophosphorus and carbamate pesticides that are broadly used in farming. They are highly toxic to humans and their residues in foods carry latent threat to human health and life. [72]

3. Third is Chronic toxicity is lethal but are more commonly sublethal, including changes in growth, reproduction, or behavior and have an effect on organs. These pesticides also affect aquatic creatures. Other examples of Chronic toxicity is drinking too much liquor and smoking tobacco.

4. Forth are known Carcinogens from caused from pesticides have been "evaluated for carcinogenicity, 69 were classified by United States EPA as either probable or likely to be human carcinogens and 99 were classified as possible human carcinogens, totaling 168 potential carcinogens." [73] Studies on these carcinogens take years to complete and to understand the effect they have on humans as well as the environment overall.

5. In fifth place developmental toxicity is when the pesticide affects the development of the fetus and or embryo.

6. Lastly, Genetic toxicity starts from damaged DNA also known as mutagenesis or gene mutation. These mutations occur in three forms chemical, physical or biological. [74]

Glyphosate-based herbicides (GBH) commonly known by its trade name Roundup are a group of pesticides that are widely used in agriculture to control weeds. They were first introduced in the 1970s and have since become the most commonly used herbicides in the world. Despite their widespread use, concerns

72 https://pubmed.ncbi.nlm.nih.gov/24907623/
73 https://www.ncbi.nlm.nih.gov/pmc/articles/PMC7881680/
74 https://www.toxmsdt.com/33-systemic-toxic-effects.html

have been raised about the potential health risks associated with glyphosate exposure. In addition to being present in fields, glyphosate can persist in soil for several months and can be detected in various environmental media, including plants, soil, water, and food products. This has led to the establishment of maximum residue limits (MRLs) for glyphosate residues in foods, several samples are up to 5,000 times higher than allowed.

However, there are concerns that these limits may not be protective of human health, particularly in light of recent studies that have detected glyphosate residues in food products at levels far exceeding the established MRLs. Moreover, the effects of exposure to multiple pesticides, herbicides, and fungicides, which are often mixed together and applied to crops such as soybeans and corn, are not well understood. This underscores the need for further research to better understand the potential health risks associated with glyphosate exposure and the use of GBHs in agriculture. [75]

Glyphosate can persist in the soil for several months after application, and studies have detected the presence of glyphosate residues in food samples, including cereal grains, soybeans, and baby food. The toxicity of glyphosate is a matter of debate, with conflicting studies suggesting both potential harm and safety. Nonetheless, the use of glyphosate has been banned or restricted in several countries, and its safety continues to be a subject of scrutiny.

In the national Library of Medicine November 2021 article Glyphosate Use, Toxicity and Occurrence in Food analysis of one chemical glyphosate is applied in fields and can last several months in soil raises concerns about the impact it presents in food that can cause in humans. This study shows the toxicity occurrence in food samples and violative levels. Violative levels far exceed maximum residue limits (MRLs) established for glyphosate residues in foods. [76]

---

75 https://www.ncbi.nlm.nih.gov/pmc/articles/PMC8622992/
76 https://www.ncbi.nlm.nih.gov/pmc/articles/PMC8622992/

Moreover, the use of multiple chemicals on crops is a common practice in modern agriculture. Studies have shown that farms can apply as many as 87 different chemicals on a single crop in a season, raising concerns about the residual effect of these chemicals on the environment and human health. With respect to researchers, we do not know the residual effect to animals, plants, insects' amphibians, or humans when multiple chemicals are mixed and applied to fields. [77] This interaction between different chemicals can create new toxic compounds, and the long-term effects of these compounds are not well understood. The use of pesticides in agriculture is a complex issue that requires further research to fully understand the impact on the environment and human health.

Studies show only a lingering effect of glyphosate in food. The European Food Safety Authority study from 2015 has inconsistent results. Along with conflicting reports from the EPA stating glyphosate is not a carcinogen, but the Agency for Toxic Substances and Disease Registry declared glyphosate is a potential carcinogen. Therefore, there is no consensus as to glyphosate being a potential carcinogen.

Despite the EPA's conclusion that glyphosate is not a carcinogen, some experts and organizations have raised concerns about its potential health impacts. The International Agency for Research on Cancer (IARC) classified glyphosate as a "probable human carcinogen" in 2015, which contradicts the EPA's position. [78] The IARC's decision was based on a review of numerous studies, including both animal and human studies, which suggested a possible link between glyphosate and certain types of cancer.

Moreover, in 2021, the Environmental Working Group (EWG), a non-profit environmental organization, released a report claiming that almost all samples of oat-based cereals, snack bars, and

---

77 https://www.nass.usda.gov/Surveys/Guide_to_NASS_Surveys/Chemical_Use/2019_Field_Crops/chem-highlights-wheat-2019.pdf
78 https://www.reuters.com/investigates/special-report/who-iarc-glyphosate/

granola tested positive for traces of glyphosate, even though the levels detected were below the EPA's legal limit. [79]

The EWG argued that long-term exposure to even low levels of glyphosate could be harmful to human health, especially for children. Despite the conflicting reports and opinions on glyphosate's potential health risks, it is clear that more research is needed to fully understand the effects of long-term exposure to this chemical on human health and the environment.

Food and Drug Administration (FDA) is responsible under the Federal Food, Drug, and Cosmetic Act for enforcing tolerances established by the Environmental Protection Agency (EPA) for amounts of pesticide residues that may legally remain on food (including animal feed). 80

There are recommendations about mixing pesticides, herbicides and fungicides and recommends there are benefits for some insects to survive. 81 This recommendation did not suggest there could be chemical consequences unknowns by mixing pesticides, herbicides, and fungicides together.

Anaphylactic shock

In 2014, a case involving a 10-year-old girl from Illinois brought attention to the potential dangers of using antibiotic pesticides on fruits and vegetables. The girl experienced anaphylactic shock after consuming blueberry pie, and doctors initially struggled to determine the cause of her severe allergic reaction. Eventually, it was discovered that the blueberries used in the pie had been treated with an antibiotic pesticide called streptomycin, marking the first reported case linking an allergic reaction to fruits treated with antibiotic pesticides [82].

79 https://www.ewg.org/news-insights/news-release/2018/10/roundup-breakfast-part-2-new-tests-weed-killer-found-all-kids
80 https://www.fda.gov/food/chemical-contaminants-pesticides/pesticides
81 https://www.canolacouncil.org/canola-watch/2018/05/30/tank-mixes-of-herbi-cides-and-insecticides/
82 https://modernfarmer.com/2014/09/10-year-old-girl-severe-allergic-reaction-pes-ticides-blueberry-pie/

Antibiotic pesticides are commonly employed in agriculture to control bacterial diseases, but they have been associated with various health and environmental concerns. In the United States, there are currently more than 4,000 antimicrobial products registered with the Environmental Protection Agency (EPA) and available in the market [2] [83]. Contrastingly, in Europe, the use of streptomycin and other antibiotic pesticides has been prohibited since 2004 [84].

This case highlights the need for greater awareness and regulation of antibiotic pesticides in the US. It also underscores the importance of organic farming practices, which prohibit the use of synthetic pesticides and antibiotics. By choosing organic produce, consumers can reduce their exposure to potentially harmful chemicals and support more sustainable and environmentally friendly farming practices. There are more than 4,000 antimicrobial products that are currently registered with EPA and sold in the open market. [85]

It is important to note that the pesticide industry still faces challenges in terms of regulation. While certain regulations are in place, the possibility of inadvertently mixing different pesticides remains a concern. Pesticides are typically tested individually in laboratories, and the practice of combining multiple pesticides in a single vat is not common.

Choose organic. Reduce harmful exposure to chemicals.

---

83 https://www.epa.gov/pesticide-registration/information-antimicrobial-products
84 https://eur-lex.europa.eu/legal-content/EN/TXT/?uri=CELEX:32005R0396
85 https://modernfarmer.com/2014/09/10-year-old-girl-severe-allergic-reaction-pesticides-blueberry-pie/

GMO

GMOs are a controversial topic in the American food system. Some argue that genetically modified crops can increase crop yield, lower the cost of food production, and help to combat global hunger. However, others argue that GMOs have potential risks to human health and the environment.

One of the main concerns about GMOs is the use of genetically modified seeds. Large agricultural companies sell GMO seeds to farmers, who then plant these seeds to grow crops. Once these crops are harvested, they are often used to make ingredients such as cornstarch, corn syrup, soybean oil, and other food additives that are present in many of the processed foods Americans consume.

Critics argue that genetically modified seeds may have unintended consequences. For example, the use of GMO seeds can lead to the development of superweeds and pests that are resistant to traditional herbicides and pesticides. Additionally, there are concerns about the potential long-term effects of consuming genetically modified foods on human health.

Despite these concerns, many crops in the United States are genetically modified, including corn, soybeans, and cotton. The USDA has approved the use of GMO technology in a variety of crops, including alfalfa, apple varieties, eggplant varieties, papaya, pineapple, potato, salmon, squash, and sugar beet. While GMO crops are primarily used for animal feed, they are also used in human food products.[86]

Pesticides have become a common feature in our food system, and their use is not limited to agriculture and food production. Many people use pesticides in their homes, gardens, and even on our skin to control pests. However, the health effects of pesticides are still not well understood, and they can have potential negative

---

86 https://www.fda.gov/food/agricultural-biotechnology/gmo-crops-animal-food-and-beyond

impacts on the nervous, endocrine, and reproductive systems.[87] The rise of GMOs has changed the landscape of pesticide usage in crop fields. Many GMOs are designed to be resistant to pesticides or produce pesticides themselves, allowing farmers to apply higher doses of pesticides without damaging their crops. This has led to an increase in pesticide use and exposure, as well as concerns about the potential health effects of consuming foods treated with these chemicals.

Moreover, there is a lack of long-term studies on the safety of GMOs and their impact on human health and the environment. While some argue that GMOs offer a solution to world hunger and food insecurity, others are concerned about the potential risks and the need for more research and regulation. As such, it is important to carefully consider the risks and benefits of pesticide use and GMOs in our food system, and to make informed choices about what we consume. [88]

Another concern with GMOs is their impact on biodiversity. GMOs have the potential to displace traditional crops and plant varieties, leading to reduced biodiversity and genetic diversity. This can have negative effects on the environment and our food systems. Additionally, there are concerns about the long-term effects of consuming GMOs, as their impact on human health is still not fully understood. Some studies have suggested that GMOs may have negative effects on gut bacteria, which are important for overall health and immune function. [89]

Furthermore, the use of GMOs and pesticides can lead to the development of pesticide-resistant insects and weeds, which can cause a need for more pesticides to be used, leading to a vicious cycle. This not only increases the potential for harm to human health but can also have negative impacts on the environment, such as water and soil contamination.

---

87 https://sitn.hms.harvard.edu/flash/2015/gmos-and-pesticides/
88 https://sitn.hms.harvard.edu/flash/2015/gmos-and-pesticides/
89 https://www.medicalnewstoday.com/articles/324576#cons

The lack of transparency and regulation in the GMO industry also raises questions about the potential for conflicts of interest and the prioritization of profit over public health and environmental concerns. It is important for consumers to have access to accurate and transparent information about the products they consume, and for there to be rigorous testing and regulation of GMOs to ensure their safety and minimize potential harm to human health and the environment.

To expand on the hidden facts about GMO foods, there have been studies that suggest that they can cause toxic effects on the body. For example, some studies have found that GMOs may cause hepatic, pancreatic, renal, or other reproductive issues. Additionally, GMOs could potentially alter hematological, biochemical, and immunologic parameters, which are important for overall health and wellbeing. [90]

However, it's worth noting that not all studies have found negative effects of GMOs on human health, and there is ongoing debate among scientists and researchers about the safety of GMOs. Despite this, there are concerns among some consumers about the potential risks associated with consuming GMO foods, and some countries have implemented regulations or labeling requirements to address these concerns.

Overall, the lack of transparency and clear information about GMOs can be a source of concern for many consumers, particularly those who prioritize knowing what they are consuming and how it may affect their health. [91]

The study mentioned raises important questions about the long-term effects of consuming genetically modified organisms (GMOs), especially when it comes to the potential for liver and other organ damage. The liver is one of the most important organs in the body, responsible for filtering out toxins and producing bile, which helps with digestion. If the liver is damaged, it can lead to a range of health problems.

90 https://pubmed.ncbi.nlm.nih.gov/18989835/
91 https://www.tandfonline.com/doi/abs/10.1080/10408390701855993

In the case of MON863, a form of GMO corn produced by Monsanto, the study found that rats fed this corn showed signs of liver and kidney damage. This is significant because MON863 was authorized for human consumption and is one of many GMO crops used in the food supply chain. The study highlights the need for further research into the potential health effects of GMOs and the importance of transparency in the food industry. Consumers have a right to know what they are eating, and the potential risks associated with consuming genetically modified foods. [92]

There have been some studies on the effects of GMO soybeans on mice, including one study that found that mice fed with GMO soybeans had disturbed pancreatic microstructures. Specifically, the study found that the mice had abnormal cell growth in the pancreas, which could potentially lead to pancreatic damage or disease. However, it is important to note that more research is needed to fully understand the effects of GMOs on human and animal health. [93]

Manuela Malatesta was an Italian scientist who conducted research on the health effects of GMOs in 2003. She published two groundbreaking studies on the topic but faced backlash and financial difficulties as a result of her findings. Unfortunately, she was unable to complete further studies and eventually was fired from her position.

Another study was conducted by a team led by Calderón de la Barca in Mexico, which found that GMO soy protein isolate had harmful effects on the pancreas. This study was conducted over a shorter duration than Malatesta's studies but added to the growing body of evidence indicating potential health risks associated with GMOs. [94]

---

92 https://www.reuters.com/article/us-monsanto-greenpeace/gmo-corn-causes-liver-kidney-problems-in-rats-study-idUSL1346840020070313

93 https://www.researchgate.net/publication/5805598_Pancreatic_response_of_rats_fed_genetically_modified_soybean

94 https://www.gmwatch.org/en/106-news/latest-news/20129-gm-soy-injures-the-pancreas-rat-feeding-study-shows

In addition to corn, soy is another major GM crop to avoid. Soy is used in many food products including soybean oil, soy flour, soy lecithin, and soy protein isolate. Soy is also commonly used as an ingredient in processed foods, such as snack bars, breakfast cereals, and vegetarian meat substitutes. However, many studies have shown that GMO soy can cause various health concerns, including allergic reactions, hormonal imbalances, and reproductive issues.

Other GM foods to avoid include canola oil, which is commonly used in processed foods and salad dressings, and cottonseed oil, which is often used in cooking oils, margarine, and shortenings. Tomatoes and potatoes have also been genetically modified, and although they are not as prevalent as GM corn and soy, they should still be avoided if possible.

It's important to note that not all GMOs are harmful, and some GMO crops have been modified to have beneficial traits, such as being more resistant to pests or able to grow in harsher conditions. However, due to the lack of transparency in the industry and the potential health risks associated with GM foods, it's important for consumers to educate themselves and make informed choices about what they eat. Choosing organic, non-GMO options is a good way to minimize exposure to potentially harmful GM foods. [95]

Mono and diglycerides are commonly used as emulsifiers in various food products such as baked goods, ice cream, margarine, and peanut butter. These compounds are used to improve texture, consistency, and shelf life of food products. They are derived from plant and animal sources and are generally considered safe for consumption by the FDA. However, some concerns have been raised about their safety as they may contain small amounts of trans fats. Trans fats have been associated with an increased risk of heart disease, and therefore, it is recommended to limit their consumption. It is important to note that not all

---

95 https://natural-fertility-info.com/gmo-corn-hormonal-imbalance.html

mono and diglycerides contain trans fats, and the labeling of trans fats in food products has been mandatory since 2006 in the United States. If you have concerns about the use of mono and diglycerides in your diet, it is recommended to read food labels and choose products with minimal additives. [96]

To expand on the topic of plant oils being potentially GMO, it is important to note that GMO crops, such as soybean, canola, and cottonseed, are often used to produce these oils. This means that if these oils are not labeled as organic, they may contain traces of GMOs, which could be a concern for those with food allergies or other health issues.

In addition to GMOs, some plant oils may also be treated with harmful pesticides and chemicals during the cultivation process, which can potentially pose health risks. This is another reason why choosing organic and non-GMO plant oils may be a safer option.

As for trans-fats, they have been shown to have negative health effects, including increasing the risk of heart disease and stroke. It is important to note that not all plant oils contain trans-fats, and those that do may be labeled as "partially hydrogenated." It is recommended to choose oils that are not hydrogenated or labeled as "trans-fat free" to avoid the negative health effects of trans-fats. [97]

Trans-fats are a type of unsaturated fat that are commonly found in processed and fried foods. They are created through a process called hydrogenation, which turns liquid oils into solid fats. Trans-fats are often used in food production because they are cheap and increase the shelf life of products. However, they are known to have negative health effects. According to the World Health Organization, approximately 540,000 deaths occur each year due to industrially produced trans-fats. Consuming high

---

96 https://www.foodingredientfacts.org/facts-on-food-ingredients/sources-of-food-in-gredients/mono-diglycerides/

97 https://www.hsph.harvard.edu/nutritionsource/2015/04/13/ask-the-expert-concerns-about-canola-oil/

amounts of trans-fats can increase the risk of death by 34%, heart disease by 28%, and coronary heart disease by 21%. In response to these health concerns, many countries have implemented policies to reduce or eliminate the use of trans-fats in food production. [98]

Vegetable oil is a commonly used ingredient in many processed foods, and it is often sourced from genetically modified crops such as corn, soybean, cotton, and canola. These crops are engineered to resist herbicides and pesticides, allowing farmers to spray them with these chemicals to control weeds and insects. As a result, vegetable oils derived from these crops may contain residues of these harmful chemicals.

In addition to concerns about pesticides and herbicides, vegetable oils also have a high concentration of omega-6 fatty acids, which can contribute to inflammation in the body when consumed in excess. The average Western diet tends to be high in omega-6 fatty acids, which can disrupt the delicate balance between omega-3 and omega-6 fatty acids and contribute to a range of chronic diseases, including heart disease, cancer, and autoimmune disorders.

It is important to note that not all vegetable oils are created equal, and some may be healthier than others. For example, oils like olive oil and avocado oil are rich in healthy monounsaturated and polyunsaturated fats, while also providing important antioxidants and other beneficial nutrients. When choosing vegetable oil, it is important to look for oils that are cold-pressed and unrefined, as these will typically retain more of their natural nutrients and flavor.

Polysorbate 80 is a common food additive used to emulsify and stabilize processed foods, such as ice cream, salad dressings, and baked goods. While it is generally considered safe in small quantities by the FDA, concerns have been raised about its potential health effects.

---

98 https://www.who.int/news-room/questions-and-answers/item/nutrition-trans-fat

One concern is that Polysorbate 80 may increase the risk of infertility and ovary deformities in animals. [99] Some studies have suggested that it may disrupt the hormonal balance in the body, leading to these effects.

Additionally, research has found that Polysorbate 80 may increase the risk of blood clots, stroke, and heart failure in patients who are already at risk. This is because the additive has been shown to have pro-inflammatory properties, which can promote the development of these conditions.

Finally, there is some evidence to suggest that Polysorbate 80 may increase the risk of tumor growth in cancer patients. While the exact mechanisms behind this are not fully understood, some researchers believe that the additive may promote the growth and spread of cancer cells.

Overall, while Polysorbate 80 is generally considered safe in small quantities, consumers may wish to limit their intake of this additive and look for products that do not contain it.

Polysorbate 80 is an emulsifying agent made of corn, palm oil, and petroleum that does not ever spoil. [100] It's been linked to infertility and ovary deformities in animals. It's also shown to increase the risk of blood clots, stroke, and heart failure in patients at risk. Also, it may increase the risk of tumor growth in cancer patients.

Powdered cellulose is a white, odorless, and tasteless powder that is used as a food additive and is commonly found in processed foods. It is made from plant fibers that are processed and refined, often from wood pulp or cotton. While it is generally considered safe for human consumption, some people may experience digestive issues such as bloating, gas, and constipation when consuming large amounts of powdered cellulose.

In addition to its use as an anti-caking agent for shredded cheese, powdered cellulose is also used as a thickener, stabilizer, and emulsifier in many other food products, including baked goods, processed meats, and canned fruits and vegetables. While

---

99   https://www.ncbi.nlm.nih.gov/pmc/articles/PMC4528880/
100 https://thegoodhuman.com/what-is-polysorbate-80/

it is derived from plant material, the processing involved in creating powdered cellulose can involve the use of various chemicals, such as sodium hydroxide and sulfuric acid.

As with any processed food additive, it is important to read food labels and be aware of the ingredients in the foods you consume. If you are concerned about the use of powdered cellulose in your food, consider opting for whole foods and minimally processed foods whenever possible.

Soy lecithin is a byproduct of the processing of raw soybean oil known as de-gumming. [101] During this process, the crude oil is treated with water, acid, or steam to remove the gums, which are natural emulsifiers that can cause cloudiness and instability in the oil. The resulting sludge is then separated from the oil, dried, and processed into a yellowish-brown, oily substance known as soy lecithin.

However, soy lecithin is not just a harmless byproduct. Since it is derived from the waste material of soybean oil production, it can contain a number of unwanted contaminants. The de-gumming process often involves the use of harsh chemicals such as hexane, a petroleum-based solvent, which can leave behind residues in the final product. In addition, soybean crops are often heavily sprayed with pesticides, which can also find their way into the soy lecithin.

Therefore, soy lecithin may contain chemical solvents and pesticide residues, making it a potential health hazard. Although the amounts of these contaminants are generally considered to be small, some people may be more sensitive to their effects. For example, individuals with soy allergies or sensitivities may experience adverse reactions to soy lecithin.

Overall, while soy lecithin is commonly used as an emulsifier and stabilizer in processed foods, it's important to be aware of its potential risks and to consume it in moderation.

Soy lecithin is the sludge leftover when raw soybean oil has gone through the "de-gumming" process. As a waste product, it's

---

101 https://www.lecitein.com/blog/all-you-need-to-know-about-degumming-of-soy-lecithin

filled with chemical solvents used in the manufacturing process and pesticides from the agricultural fields.

Do you know what's in your diary? Here's a list of 30 dairy additives with an easy-to-understand explanation as to what they all are. Know what's in your food before you buy it!:: DontWasteth-eCrumbs.com

Sucralose is an artificial sweetener, commonly found in little yellow packets. Also known as Splenda. Sucralose has been proven to cause several serious medical conditions including seizures, dizziness, migraines, blurred vision, and gastrointestinal problems. It also increases blood sugar and causes weight gain.

Hydrogen peroxide (H_2O_2) has been used for decades as a preservative in milk around the world. The main reason for its use is its ability to activate the inherent lactoperoxidase enzyme system present in raw milk. Lactoperoxidase is a naturally occurring enzyme found in milk, saliva, and other bodily fluids that plays an important role in the nonspecific immune response involved in maintaining oral health. Its main function is to catalyze the oxidation of thiocyanate ions ($SCN-$) present in saliva in the presence of hydrogen peroxide (H_2O_2) to generate reactive intermediates that exhibit antimicrobial activity against a range of microorganisms.

The addition of hydrogen peroxide to raw milk enhances the antimicrobial activity of lactoperoxidase, which in turn improves the quality and shelf-life of the milk. This method is particularly important in areas where refrigeration and other cooling methods are not widely available, as it provides a means of preserving raw milk for longer periods of time. The use of hydrogen peroxide also helps to reduce bacterial contamination in raw milk by inhibiting the growth of spoilage and pathogenic microorganisms.

Moreover, the use of hydrogen peroxide is a cost-effective and safe way to preserve milk compared to other chemical preservatives. The amount of hydrogen peroxide added to milk is carefully regulated to ensure that it does not exceed the recommended limit of 20 mg/L, which is considered safe for human consumption. The residual levels of hydrogen peroxide in milk after treatment are typically very low and do not pose any health risks to consumers.

In conclusion, the use of hydrogen peroxide as a preservative in milk has a long-established history and is an effective means of preserving raw milk in areas where cooling methods are limited. The activation of the lactoperoxidase enzyme system by hydrogen peroxide enhances the antimicrobial activity of raw milk and improves its quality and shelf-life. The use of hydrogen peroxide is safe, cost-effective, and does not pose any health risks to consumers when used in accordance with recommended guidelines. [102]

Sugar beets are a major source of sugar production worldwide, with over a hundred million tons produced annually. However, the use of pesticides and herbicides is common in sugar beet farming to control weed growth and pests, which can negatively impact the quality of the crop and reduce yields. Unfortunately, these chemicals can also have detrimental effects on human health and the environment.

To address this issue, some people are turning to organic honey as a substitute for sugar beets in their diets. Organic honey is a natural sweetener that is produced by bees from the nectar of flowers. Unlike sugar beets, honey does not require any pesticides or herbicides to be applied to the crop, as bees collect nectar from a wide range of plants in their natural environment.

Organic honey also has a number of health benefits compared to sugar beets. It contains antioxidants, enzymes, and other beneficial compounds that can help to boost the immune system and support overall health. Honey is also a good source of energy and can be used as a natural remedy for sore throats and coughs.

In addition, using organic honey as a substitute for sugar beets can have positive environmental impacts. Honeybees play a crucial role in pollinating crops and plants, which helps to maintain biodiversity and support ecosystems. By supporting the produc-

---

[102] https://dontwastethecrumbs.com/30-additives-in-dairy-products-you-should-know-about/

tion and consumption of organic honey, we can help to promote sustainable agricultural practices and reduce our reliance on pesticides and herbicides.

Overall, while sugar beets remain a major source of sugar production worldwide, the use of pesticides and herbicides in their cultivation can have negative impacts on human health and the environment. [103] Using organic honey as a substitute for sugar beets can provide a natural and healthy alternative that supports sustainable agriculture and promotes overall well-being.

Soybeans are one of the most widely cultivated crops in the world and are used to produce a variety of food products including soy milk, tofu, soy oil, soy sauce, and infant formulas, among others. Soybeans that have had pesticides or herbicide applied has been linked to pancreatic issues. [104]

However, a significant portion of soybeans grown today are genetically modified (GM) or genetically engineered to be resistant to herbicides and pesticides. [105] These GM soybeans have been developed to increase yields and reduce the need for chemical inputs, but concerns have been raised about their potential impact on human health.

One of the main issues associated with GM soybeans is their potential link to pancreatic issues. [106] Studies have shown that the consumption of soybeans that have been treated with pesticides or herbicides can increase the risk of pancreatic issues, including cancer. This is because these chemicals can disrupt the normal functioning of the pancreas, which plays a crucial role in regulating blood sugar levels and producing digestive enzymes. [107] Furthermore, the widespread use of GM soybeans has raised concerns about the long-term health effects of consuming these products. While some studies have shown that GM soybeans are safe for human consumption, others have raised concerns about

103 https://onlinelibrary.wiley.com/doi/10.1111/j.1467-7652.2004.00076.x
104 https://www.ncbi.nlm.nih.gov/pmc/articles/PMC7530464/
105 https://www.fda.gov/food/agricultural-biotechnology/gmo-crops-animal-food-and-beyond
106 https://www.ncbi.nlm.nih.gov/pmc/articles/PMC1568217/
107 https://sitn.hms.harvard.edu/flash/2015/will-gmos-hurt-my-body/

their potential impact on human health, including their potential to trigger allergic reactions and affect hormonal balance.

It is worth noting, however, that not all soybeans are genetically modified or treated with pesticides or herbicides. Organic soybeans, for example, are grown without the use of these chemicals, and are a safer and healthier option for consumers. [108] In addition, there are many non-GMO soy-based products available on the market that are made from soybeans that have not been genetically modified. [109]

In conclusion, while soybeans are a versatile and widely used crop, the use of GM soybeans and the application of pesticides and herbicides to soybean n crops have been linked to potential health issues, including pancreatic issues. To minimize these risks, it is important to choose organic and non-GMO soy-based products, and to support sustainable agricultural practices that prioritize human health and environmental well-being.

Genetically modified organisms (GMOs) have become a popular topic of debate in recent years, particularly in the context of food production. One common GMO crop is the tomato, which has been genetically engineered to be resistant to pests and to have a longer shelf life. However, research has shown that GMO tomatoes may have lower antioxidant levels compared to their organic counterparts.

Antioxidants are important nutrients that help to protect the body against damage from free radicals, which are unstable molecules that can cause oxidative stress and contribute to chronic diseases like cancer, diabetes, and heart disease. Organic tomatoes have been found to contain higher levels of antioxidants compared to GM tomatoes, likely because organic farming practices prioritize soil health and biodiversity, which can increase the availability of nutrients in the soil.

In addition, some studies have suggested that GMO tomatoes may have negative impacts on human health. For example, one study found that rats fed a diet containing GMO tomatoes

---

108 https://www.ota.com/health-benefits-organic
109 https://www.non-gmoreport.com/articles/jul08/non-gmo_soybean_seed.php

experienced adverse effects on their liver and kidneys, as well as changes in their blood chemistry. [110] While these findings are not conclusive, they raise concerns about the potential risks associated with consuming GMO tomatoes.

Despite these concerns, GMO tomatoes remain a widely cultivated crop, particularly in areas where pest resistance and shelf life are important factors for economic viability. However, consumers can choose to prioritize organic tomatoes and other organic produce, which are grown without the use of GMOs or synthetic pesticides and are often higher in nutrients like antioxidants.

In conclusion, while GMO tomatoes offer certain benefits like pest resistance and longer shelf life, research has shown that they may have lower antioxidant levels compared to organic tomatoes, which can have negative impacts on human health. [111] To minimize these risks, consumers can choose to prioritize organic produce and support sustainable agricultural practices that prioritize soil health and biodiversity.

Beef is a popular protein source consumed by many people around the world. However, the quality of beef can vary depending on a number of factors, including the diet of the animals from which it is sourced. In recent years, concerns have been raised about the use of genetically modified organisms (GMOs) in livestock feed, particularly in the context of mono farms. [112]

A mono farm is a type of industrial farming operation in which a single crop is grown over a large area, often with the use of synthetic fertilizers and pesticides. Many mono farms grow crops like alfalfa, corn, and soybeans, which are commonly used as feed for cattle. However, a significant portion of these crops are genetically modified or genetically engineered to be resistant to pests and herbicides.

This means that beef sourced from mono farms that are not organic may contain GMO or GM alfalfa, corn, or soybeans. While there is some debate about the safety of consuming GMOs, many

---

110 https://www.ncbi.nlm.nih.gov/books/NBK83984.7/?report=reader
111 https://www.ncbi.nlm.nih.gov/pmc/articles/PMC8869745/
112 https://www.sciencedirect.com/science/article/abs/pii/S0278691517304842

consumers are concerned about their potential impact on human health and the environment. Some studies have suggested that GMOs may have negative impacts on human health, including increased risk of certain cancers, hormonal imbalances, and digestive issues.

Furthermore, the use of GMOs in livestock feed has raised concerns about animal welfare, as well as the potential environmental impact of mono farming operations. For example, the use of synthetic fertilizers and pesticides can contribute to soil degradation and water pollution, which can have negative impacts on both human and animal health.

To minimize these risks, consumers can choose to prioritize beef sourced from organic farms, which are required to meet certain standards regarding animal welfare, environmental sustainability, and the use of GMOs and synthetic chemicals. [113] In addition, consumers can support sustainable farming practices like regenerative agriculture, which prioritize soil health and biodiversity, and can help to reduce the use of synthetic chemicals and the reliance on monoculture crops.

In conclusion, beef sourced from mono farms that are not organic may contain GMO or GM alfalfa, corn, or soybeans, which can have negative impacts on both human health and the environment. [114] To minimize these risks, consumers can choose to prioritize organic beef and support sustainable farming practices that prioritize animal welfare, environmental sustainability, and the use of natural and organic feed sources.

Cows are commonly fed GMO or GM alfalfa, corn, or soybeans in order to increase milk output. In addition to this, some dairy farmers also use bovine growth hormone (rGBH) to further increase milk production. [115] rGBH is also linked to early breast cancer in women. However, this practice has been controversial due to concerns about its impact on both animal and human health.

---

113 https://www.organic-center.org/site/benefits-organic-meat
114 https://www.frontiersin.org/articles/10.3389/fenvs.2021.763917/full
115 https://www.bcpp.org/resource/rbgh-rbst/

Studies have shown that milk from cows injected with rGBH may contain reduced amounts of healthy fatty acids, such as conjugated linoleic acid (CLA), which has been linked to a range of health benefits including reduced inflammation and improved cardiovascular health. In addition, the use of GMO or GM feed may also contribute to the presence of potential pesticides in the milk, as cows are exposed to these chemicals when they consume feed that has been treated with pesticides.

Furthermore, there are concerns about the impact of rGBH on the welfare of dairy cows. The use of this hormone can lead to increased stress and health problems, as cows are pushed to produce more milk than they would naturally. This can lead to issues such as mastitis, a painful inflammation of the udder that can cause significant discomfort for cows.

To address these concerns, many consumers choose to prioritize organic dairy products, which are produced without the use of GMO or GM feed, synthetic hormones, or antibiotics. Organic farming practices prioritize animal welfare, environmental sustainability, and the use of natural and organic feed sources, which can help to reduce the use of synthetic chemicals and the reliance on monoculture crops.

In conclusion, cows that are fed GMO or GM alfalfa, corn, or soybeans and injected with bovine growth hormone (rGBH) may produce milk with reduced amounts of healthy fatty acids and potential pesticides from their feed. To minimize these risks, consumers can choose to prioritize organic dairy products and support sustainable farming practices that prioritize animal welfare, environmental sustainability, and the use of natural and organic feed sources.[116]

---

116 https://explore.globalhealing.com/top-20-gmo-foods-and-ingredients-to-avoid/

There is growing concern about the potential risks associated with gene transfer between genetically modified organisms (GMOs) and wild populations. In particular, the transfer of pesticide, herbicide, or antibiotic resistance genes could have serious ecological consequences, promoting the spread of disease in plants and animals and causing imbalances in natural ecosystems.

However, studying the effects of gene transfer between GMOs and wild populations is a complex and challenging task, as it cannot be easily duplicated in a laboratory setting. This presents a significant challenge when attempting to evaluate the potential risks associated with GMOs, as chemical companies may argue that these risks have not been adequately demonstrated.

One area of research that has shed light on the potential risks associated with GMOs is the study of transgenic fish. [117] These are fish whose genomes have been altered by one or more foreign DNA sequences, which are then released into wild populations of the same species. [118] Studies have found that transgenic fish can have mating advantages over their non-GMO counterparts, leading to a reduction in the viability of their offspring and potentially threatening the viability of the original species.

This research highlights the need for caution when introducing GMOs into the natural environment. While these organisms may offer potential benefits in terms of increased crop yields, reduced pesticide use, or improved disease resistance, the risks associated with gene transfer must be carefully evaluated to avoid unintended consequences.

In conclusion, gene transfer between GMOs and wild populations could have serious ecological consequences, promoting the spread of disease and causing imbalances in natural ecosystems. The study of transgenic fish has highlighted some of the potential risks associated with GMOs, including reduced offspring viability

---

117 https://www.sciencedaily.com/releases/2009/08/090827073250.htm
118 https://www.ncbi.nlm.nih.gov/pmc/articles/PMC9818732/

and threats to the viability of the original species. To minimize these risks, careful evaluation of GMOs is necessary before they are introduced into the natural environment. [119]

Over the past two decades, a growing body of scientific research has highlighted the potential health risks associated with GMOs and pesticides. Numerous studies have shown that exposure to these substances can have negative impacts on human health, including increased risk of cancer, reproductive problems, and neurological disorders.

Despite the significant amount of evidence supporting these findings, there remains considerable debate within the scientific community and among policymakers about the risks associated with GMOs and pesticides. Critics of this research often point to studies that show little or no evidence of harm, arguing that the risks have been overstated or that the studies are flawed. [120]

However, a close examination of the evidence suggests that the weight of the evidence supports the conclusion that pesticides, herbicides, and GMOs are indeed harmful to human health. While individual studies may be subject to critique and interpretation, the collective body of evidence is highly conclusive.

It is important to note that the vast amount of scientific literature on this topic can be difficult to navigate, particularly for non-experts. However, there are a number of resources available that summarize the key findings of these studies in accessible language. This information can be used by policymakers and the general public to make informed decisions about the use of GMOs and pesticides in agriculture and other contexts.

In conclusion, over the past twenty years, scientists have issued numerous warnings about the potential health risks associated with GMOs and pesticides. While some continue to question the evidence, the collective body of scientific literature is highly conclusive that these substances can have negative impacts on

---

119 https://www.nature.com/scitable/topicpage/genetically-modified-organisms-gmos-transgenic-crops-and-732/
120 https://pubmed.ncbi.nlm.nih.gov/26767435/

human health. As such, policymakers and the public should carefully consider this evidence when making decisions about the use of these substances.

The consolidation of production capabilities into geographic areas is a common practice among many industries, including agriculture. This strategy allows companies to streamline their production processes and take advantage of economies of scale, which can help reduce costs and increase efficiency. However, it can also lead to the concentration of animals and crops in specific regions, which can have negative environmental impacts such as soil erosion, water pollution, and loss of biodiversity.

Furthermore, the practice of feeding chickens with corn and soybean meal, along with vitamin and mineral supplements, is common in the poultry industry. Corn and soybeans are commonly used because they are a cheap source of protein and carbohydrates. However, as mentioned before, these crops are often genetically modified, and some studies have linked the consumption of GMOs with negative health effects in humans.

Additionally, some farmers use antibiotics to promote growth and prevent disease in their chickens, which has raised concerns about antibiotic resistance in humans. Antibiotic-resistant infections are becoming increasingly common and can be difficult to treat, leading to increased healthcare costs and mortality rates.

Overall, the consolidation of production capabilities and the use of GMOs and antibiotics in the poultry industry have raised concerns about environmental sustainability and public health. It is important for companies and governments to prioritize sustainable and ethical practices in food production to ensure the health and well-being of both humans and the environment.

The cattle industry in the western states of Oregon, Idaho, Montana, Wyoming, Arizona, Colorado, South Dakota, New Mexico, and Nebraska are fighting grasshoppers because of climate change as of 2023. We cannot spray enough pesticides to kill grasshoppers as it would require too many pesticides to be applied. [121]

The increased use of pesticides, herbicides, fungicides, and GMOs or GMs in the United States since the 1960s has had a significant impact on both the environment and human health. While the use of these chemicals and genetically modified crops was initially seen as a way to increase crop yields and reduce the use of harmful chemicals, it has become increasingly clear that they come with significant risks.

Studies have shown that exposure to pesticides, herbicides, and fungicides can have negative impacts on human health, including an increased risk of cancer, reproductive problems, neurological damage, and developmental delays in children. [122] Additionally, the use of these chemicals has been linked to the decline of pollinators such as bees and butterflies, which are critical for maintaining healthy ecosystems and ensuring food security. [123]

In response to these concerns, many countries around the world have banned the use of certain pesticides, herbicides, fungicides, and GMOs or GMs. For example, the European Union has banned the use of neonicotinoid pesticides, which have been linked to the decline of bee populations. [124] Additionally, many countries have strict regulations in place to ensure that genetically modified crops are safe for human consumption and do not harm the environment.

---

121 https://www.aphis.usda.gov/plant_health/ea/2023/mt-1-2023-28.pdf
122 https://www.ncbi.nlm.nih.gov/pmc/articles/PMC9428564/
123 https://www.unep.org/news-and-stories/story/why-bees-are-essential-people-and-planet
124 https://www.theguardian.com/environment/2013/apr/29/bee-harming-pesticides-banned-europe

Despite these concerns, the use of pesticides, herbicides, fungicides, and GMOs or GMs continues to be widespread in the United States. Many farmers see these chemicals and crops as essential to their livelihoods, and companies continue to produce and market them as safe and effective. However, the growing body of research on the negative impacts of these chemicals and crops has led to increasing calls for greater regulation and more sustainable agricultural practices.

The increase in grasshopper populations in the western states of the United States is primarily attributed to climate change. The warmer and drier weather conditions have created ideal breeding grounds for grasshoppers, leading to a surge in their numbers. Grasshoppers cause significant damage to crop and pastures, and the cattle industry is particularly affected by their presence. The industry has been struggling to combat grasshopper infestation as the use of pesticides is not a viable solution due to the potential environmental impact and the high cost of treating such large areas. As a result, many ranchers have been forced to find alternative methods to protect their livestock and livelihoods. Some have turned to natural predators, such as birds and insects, while others are experimenting with alternative feeds and grazing strategies to reduce their reliance on pastures. The grasshopper infestation has become a significant challenge for the cattle industry and highlights the growing impact of climate change on agriculture. [125]

The European Union banned GMO or GM [126] Encouraging news six counties in California have successfully banned GMO or GM crops in Humboldt, Mendocino, Trinity, Santa Cruz, Sonoma, and Marin as of 2022. [127]

---

125 https://www.cnn.com/2021/07/02/us/grasshoppers-cattle-drought-climate-change/index.html
126 https://ec.europa.eu/environment/europeangreencapital/countriesruleoutgmos/
127 https://gmowatch.com/where-are-gmos-banned/

GMO is not organic

Certified non-GMO foods are not the same as organic foods. Non-GMO certification only means that the food does not contain genetically modified organisms (GMOs). It does not necessarily mean that the food was grown or produced using organic methods or without pesticides or other chemicals. Therefore, it is possible for non-GMO foods to contain residues of pesticides or other synthetic chemicals.

It is also worth noting that some people may have sensitivities or allergies to specific non-GMO foods, just as they might with any other type of food. In some cases, these reactions could be due to natural compounds in the food, rather than any pesticide residue or genetic modification.

If someone is looking to avoid both GMOs and synthetic chemicals in their food, they may want to look for products that are both certified non-GMO and organic. Organic certification requires that the food was grown and processed without the use of synthetic pesticides, herbicides, and fertilizers, among other requirements.

It is important to be aware of the potential risks associated with pesticide and herbicide exposure and to take steps to minimize exposure, such as choosing organic foods or washing fruits and vegetables thoroughly before eating them.

Advice when shopping look for labels that can state GMOs, such as "genetically modified," "bioengineered," or "genetically engineered." Note labeling requirements vary by country, and not all GMOs may be labeled. [128] Or look at PLU (Price Look-Up) codes which is a five-digit PLU if the code begins with the number 8 it may indicate it is genetically modified. [129]

128 https://www.federalregister.gov/documents/2018/12/21/2018-27283/national-bio-engineered-food-disclosure-standard
129 https://www.scripps.org/news_items/4472-cracking-the-produce-sticker-code

Some packaged foods may contain GMO ingredients, such as genetically modified soybeans, corn, or canola. By reading ingredients on food packages look for common GMOs soy lecithin, corn syrup, or cottonseed oil.

And look for the non-GMO label when buying food. Currently shopping for organic food in southern California such as Vons, Trader Joes, Costco, and Walmart (super store). Frankly, none of these stores carry enough organic food products to shop at one store. With the pandemic and supply chains that have been damaged and often find the items that were purchased for months now do not exist or are on back order. Examples are some stores carry non-dairy ice cream, almond milk, while others carry organic butter, organic juices, and organic fruits and vegetables. Then in the next shopping excursion items are not available and the product is not available.

History of introduced pesticides

Human knowledge and use of pesticides goes back ten thousand years or more. Though little evidence exists today from that long ago. Based on data Sumerians used sulfur to control insects and mites. 3,200 years ago, Chinese were using insecticides made from plants and 2500 years ago humans stated adjusting crop rotation to prevent insects from harming crops. Greeks and Romans used fumigants, sticky bans around trees, placing grains on stilts, along with pesticides and ointments then after science was debunked by religion for a long-term humans use prayer. Meanwhile in China humans were using predatory ants to control pests. [130]

For example, Native Americans in North America used companion planting, where different plants are grown together in a

---

130 https://www.pitchcare.com/news-media/pesticides-a-brief-history-and-analysis.html

way that benefits each other and deters pests. [131] In medieval Europe, farmers used crop rotation to prevent soil depletion and reduce pest populations.

Sinox, which is short for sodium chlorate, was created in the late 19th century. It was initially used as an herbicide and became popular in agriculture due to its effectiveness in controlling weeds. However, Sinox was later found to be harmful to the environment and was banned in many countries. [132]

The development of modern synthetic herbicides and pesticides in the 20th century revolutionized pest control, allowing for much greater control over pests and increasing agricultural crop yields. However, the widespread use of these pesticides has also led to numerous environmental and health concerns, including pollution of soil and water, harm to non-target species, and potential harm to human health. However, as noted earlier, the use of these chemicals has also resulted in environmental and health concerns.

The widespread use of pesticides and herbicides in agriculture began in the 1940s, and by the 1960s, over 100 new chemicals had been developed and put into use. The main reason for this surge in pesticide use was the economic benefit it provided to farmers. Chemical weed control became more economically feasible than plant-disease and insect-pest control. Farmers were under pressure to produce more food to feed an increasing global population, which reached 8 billion people by November 2022.

While the use of herbicides and pesticides has been essential to modern agriculture, there have been many controversies surrounding their safety and environmental impact. For example, in the 1960s, the herbicide 2,4,5-T was found to contain dioxin, a highly toxic chemical that was associated with a wide range of health problems. As a result, many countries banned the use of

---

131 https://www.botanicalinterests.com/blogs/tips/three-sisters-gardening-native-american-companion-planting
132 The history of herbicide use for weed management on the prairies.pdf

2,4,5-T and other herbicides that contained dioxin. [133] Since then, the development of new herbicides and pesticides has continued, and the use of these chemicals has remained widespread. Glyphosate is one of the most widely used herbicides in the world and is often used in conjunction with genetically modified crops. However, glyphosate has also been associated with health and environmental concerns, and many countries have placed restrictions on its use.

The post-World War II era saw a rapid growth in the production of pesticides, with more than 100 new pesticides being created in less than two decades. One of the most significant groups of pesticides developed during this period was the phenoxy herbicides, which included 2,4-D (2,4-dichlorophenoxyacetic acid) and 2,4,5-T (2,4,5-trichlorophenoxyacetic acid). [134] These herbicides were effective in killing broad-leaved weeds and were widely used in agriculture.

However, concerns about the health and environmental impacts of 2,4,5-T led to its eventual ban in the United States in 1985. The use of 2,4-D remains widespread in many countries, including the United States, but it too has been linked to health and environmental concerns.

Another significant pesticide developed during this period was IPC (isopropyl-N-phenylcarbamate), which was used to control grass species by being applied into the soil. Like other pesticides, IPC has been linked to health and environmental impacts, and its use has declined in recent years.

Overall, the post-World War II era saw a significant increase in the production and use of pesticides, but also highlighted the need for greater scrutiny and regulation to ensure their safety and minimize their impact on human health and the environment.

2,4-D is one of the oldest and most widely available herbicides and defoliants in the world, having been commercially available since 1945, and is now produced by many chemical

---

133 https://profils-profiles.science.gc.ca/en/publication/history-herbicide-use-weed-management-prairies
134 https://www.ncbi.nlm.nih.gov/books/NBK236351/

companies since the patent on it has long since expired. It can be found in numerous commercial lawn herbicide mixtures and is widely used as a weedkiller on cereal crops, pastures, and orchards. Over 1,500 herbicide products contain 2,4-D as an active ingredient." [135] DDT (dichloro-diphenyl-trichloroethane) was also a bad herbicide invented in the 1940ies and was used for decades. In the early 1970ies countries negotiated a treaty through United Nations Environment Program, to authorize a global ban or on persistent organic pollutants (POPs) [136] Many reproductive abnormalities have been found from the use of DDT. [137] There is still a limited use of DDT for microbes that cause malaria because it still kills millions of people worldwide.

Many other organic pesticides have been developed and used in agriculture. For example, neem oil, which comes from the neem tree, is a natural pesticide that has been used for centuries in India. It works by disrupting the insect's hormonal balance and preventing them from feeding and reproducing. Other organic pesticides include pyrethrin, derived from chrysanthemum flowers, and spinosad, a fermentation product of bacteria.

Organic pesticides are often seen as a safer alternative to synthetic pesticides, as they are less harmful to humans and the environment. [138] However, they can still have negative impacts on non-target organisms and can accumulate in the environment over time. Therefore, it is important to use them in moderation and to follow proper safety protocols when applying them.

Pesticides and herbicides are chemicals used in agriculture to control pests and weeds, respectively. However, these chemicals can have harmful effects on our bodies, even at low levels of exposure. Pesticides and herbicides are known to interfere with the endocrine system, which regulates hormones in our bodies, leading to potential hormonal imbalances that can result in a variety of health issues.

---

135 https://en.wikipedia.org/wiki/2,4-Dichlorophenoxyacetic_acid
136 https://www.epa.gov/ingredients-used-pesticide-products/ddt-brief-history-and-status
137 https://www.ncbi.nlm.nih.gov/pmc/articles/PMC2726844/
138 https://hgic.clemson.edu/factsheet/less-toxic-insecticides/

For example, studies have shown that exposure to pesticides can increase the risk of certain cancers, including non-Hodgkin's lymphoma and leukemia. In addition, some pesticides have been linked to neurological disorders, such as Parkinson's disease and Alzheimer's disease. [139] Pregnant women and children are particularly vulnerable to the effects of pesticides and herbicides, as these chemicals can cross the placenta and affect fetal development. It is possible to receive a reaction to pesticides could be due to an endocrine disruption, but it is important to note that the effects of pesticides and herbicides can vary from person to person. The liver and kidneys are responsible for removing toxins from the body, but excessive exposure to these chemicals can overload the body's detoxification system and lead to negative health effects.

The corporate structure of the US food system has been designed to prioritize maximum profit and efficiency over the well-being of consumers and the environment. This often leads to practices such as factory farming, where animals are raised in confined spaces with little regard for their welfare or the health implications of their meat for human consumption. The focus on cost-cutting also means that feed costs can make up a significant portion of overall production costs, particularly for pigs and poultry. [140]

The negative effects of pesticides have become increasingly apparent, and many countries have implemented regulations to limit their use. This has led to the development of alternative farming methods, such as organic farming, which relies on natural pest control and avoids the use of synthetic chemicals. While organic farming is gaining popularity, there is still a long way to go to reduce the harmful impact of pesticides on our food system.

The use of pesticides has become a pervasive practice in the food system, affecting almost every sector of agriculture. Chemical companies have marketed pesticides to farmers, claiming that they will increase crop yields. While this may be true in the

---

139 https://www.ncbi.nlm.nih.gov/pmc/articles/PMC2231435/
140 https://www.frontiersin.org/articles/10.3389/fvets.2021.750733/full

short term, the long-term effects of these chemicals are starting to show. Pesticides have been linked to environmental damage, harm to animals, including bees, and human health problems.

Pesticides and herbicides are designed to kill organic organisms. All things living are organic. As a living organism we must have a stable environment and it is crucial for any living thing. A living thing has an organized structure, can react to stimuli, reproduce, grow, and adapt to their environment.

Spraying grains with pesticides can have a negative effect on the meat we consume. Why, because pesticides accumulate in the animals eating these grains and then pass on to humans as we consume these meat products. In addition, while spraying the crops the environment becomes contaminated adding additional health risks. [141]

In recent years, there has been a growing interest in alternative methods of pest and weed control. Integrated pest management, for example, involves the use of a variety of methods, including crop rotation, natural predators, and the careful use of pesticides, to control pests and weeds. Organic farming methods also rely on natural methods of pest and weed control and prohibit the use of synthetic herbicides and pesticides.

As the global population continues to grow, the use of herbicides and pesticides is likely to remain an essential part of modern agriculture. However, it is also essential to continue to develop safer and more environmentally friendly methods of pest and weed control to minimize the impact on human health and the environment. [142]

Paraquat is a highly toxic herbicide that has been widely used in agriculture to control weeds for decades. It is often used in the United States on crops such as cotton, soybeans, corn, and

---

141 https://www.ncbi.nlm.nih.gov/pmc/articles/PMC6422529/
142 https://www.epa.gov/ingredients-used-pesticide-products/24-d

other agricultural crops. While Paraquat is banned in many countries due to its toxicity, it continues to be used in the United States due to its effectiveness in controlling weeds. [143]

Several studies have linked exposure to Paraquat to an increased risk of Parkinson's disease, a progressive neurodegenerative disorder that affects movement and can cause tremors, stiffness, and impaired balance. In addition to the 2009 study mentioned, a 2011 study found that exposure to Paraquat increased the risk of Parkinson's disease by more than twofold. The study also found that the risk was greater in people who were exposed to the herbicide for longer periods. [144]

The mechanisms by which Paraquat may contribute to the development of Parkinson's disease are not entirely clear. However, it is believed that the herbicide may cause oxidative stress and damage to dopaminergic neurons in the brain, which are involved in the regulation of movement. Studies have also suggested that Paraquat may increase inflammation and damage to the blood-brain barrier, which could further contribute to the development of Parkinson's disease.

The use of Paraquat has become increasingly controversial in recent years, with many environmental and health advocates calling for its ban due to its toxicity and potential health risks. In 2020, the Environmental Protection Agency (EPA) announced that it would not ban the use of Paraquat in the United States but would require additional training and safety measures for those who handle and apply the herbicide. [145]

Mounting evidence suggests that exposure to Paraquat, one of the most widely used chemical herbicides in the country, has been linked to an increased risk of developing Parkinson's

---

143 https://www.nytimes.com/2016/12/20/business/paraquat-weed-killer-pesticide.html
144 https://www.nature.com/articles/cdd2009217
145 https://www.epa.gov/ingredients-used-pesticide-products/paraquat-dichloride

disease. In a 2009 study revealed that any exposure to Paraquat within 1,600 feet of a home resulted in a 75 percent increased risk of Parkinson's disease. [146]

Furthermore, according to the Centers for Disease Control and Prevention (CDC), Paraquat is a highly toxic chemical that can cause severe poisoning in case of exposure through ingestion, inhalation, and skin absorption.

Several Paraquat lawsuits have been filed across the country by plaintiffs who have been subsequently diagnosed with Parkinson's disease after being exposed to the pesticide in the workplace." [147] Paraquat is another pesticide that should be banned from being applied on crops.

Some of the countries that have prohibited Paraquat use include Brazil, Syria, Kuwait, Austria, Finland, Slovenia, Sweden, Denmark, Malaysia, Cambodia, Germany, Ivory Coast, Switzerland, United Kingdom, European Union (EU), United Arab Emirates [148] The United States is missing!

Dicamba has been described as either a benzoic acid or carboxylic acid compound, approved in 1962 used on less than 10% of US corn acres in 1979. Use increased to 15% of corn hectares by 1990, then as herbicide-resistant weeds spread, dicamba use on corn increased to 28% of hectares in 1995. From 2017 to 2019 farmers reported thousands of dicamba drift episodes causing damage to millions of acres of soybeans as well as vegetables, fruit trees, gardens, and residential trees. [149] As a group 4 pesticide it is meant to selectively control broadleaves in grass crops and is primarily used on corn and soybeans.

146 https://www.allenandallen.com/popular-herbicide-paraquat-linked-to-high-risk-of-parkinsons-disease/
147 https://www.drugwatcher.org/how-do-herbicides-affect-humans/
148 https://www.drugwatcher.org/what-countries-banned-paraquat/
149 https://biologicaldiversity.org/w/news/press-releases/federal-court-holds-dicamba-pesticide-unlawful-citing-unprecedented-drift-damage-millions-acres-2020-06-03/

Dicamba is one herbicide chemical out of hundreds of chemicals that should not be allowed to be applied on food sources. In 2020 – "The U.S. Court of Appeals for the 9th Circuit today ruled that the Trump administration wrongly approved Monsanto's pesticide dicamba for use on genetically engineered soy and cotton – a decision that makes the sale and use of the pesticide illegal." [150] This is only one additional pesticide being banned. Glyphosate has become the most used weedkiller on our planet and from 1996 about 30 million pounds were used to 2016, when 287 million pounds were applied to crops according to the U.S. Geological Survey. [151]

News of the day April 2024 reported by The Guardian news "Instead of yanking products, EPA made Monsanto and others amend labels before reapproving dicamba, lawsuit claims". "The pesticide industry has a ton of clout in the EPA's pesticide office, a ton of ability to persuade people there, and the culture at the office is very in alliance with the pesticide industry," Donley told the Guardian. [152] The Environmental Protection Agency is facing more lawsuits over herbicides, including a challenge from grower groups over new dicamba restrictions and a lawsuit from environmental groups over atrazine. [153]

Today, United States farmers spends over $11 billion on all pesticides, and 58 percent goes to herbicides, 28 percent to insecticides, 8 percent to fungicides and 6 percent to other chemicals. [154]One in three people across America have detectable levels of a toxic herbicide linked to cancers, birth defects and

---

150 https://biologicaldiversity.org/w/news/press-releases/federal-court-holds-dicamba-pesticide-unlawful-citing-unprecedented-drift-damage-millions-acres-2020-06-03/
151 https://www.salon.com/2020/03/21/how-big-corporations-are-poisoning-our-farms_partner/
152 https://www.theguardian.com/environment/2023/apr/24/epa-monsanto-toxic-herbicides-dicamba?
153 https://www.agri-pulse.com/articles/14813-epa-facing-lawsuits-over-atrazine-dicamba
154 https://livinghistoryfarm.org/farminginthe40s/pests_03.html

hormonal imbalances. Human exposure to the herbicide 2,4-D has substantially risen amid expanding use among farmers despite a multitude of health and environmental concerns, according to the first nationally representative study evaluating the footprint of the chemical.[155]

We have friends who travel to Europe and one of the things they say upon returning to the United States they say while in Europe "I could eat anything I wanted to and did not have a food reaction". We live in a time when we can fix this however, we need action from state and federal legislators and the courts to stop poising us with pesticides, herbicides, and fungicides.

According to the deal, $8.8 billion to $9.6 billion was set aside for current litigation, including a buffer amount to cover claims not resolved yet. $1.25 billion has been set aside for future claims from Roundup users with non-Hodgkin's lymphoma. Part of this amount will also be used for an independent expert panel which will decide whether glyphosate causes cancer and what exposure level, or dosage of glyphosate would be dangerous. [156]

A scientist in 1970 created Glyphosate at Monsanto's and is the main ingredient in many herbicides today. In 1996, Monsanto introduced genetically modified (GM) soybeans resistant to glyphosate.

Bayer still faces at least 30,000 claims from plaintiffs who have not agreed to join the settlement. Additionally, the company has still not admitted any liability or wrongdoing. [157]

155 https://www.theguardian.com/environment/2022/feb/09/toxic-herbicide-exposure-study-2-4-d
156 https://www.nytimes.com/2020/06/24/business/roundup-settlement-lawsuits.html
157 https://www.roundupsettlements.com/?utm_source=Google&utm_medium=PPC&utm_campaign=12079568058&utm_content=mixed&PN=8009168711&gclid=CjwKCAiArNOeBhAHEiwAze_nKAAN2uFETI-jcdKKxjZzU_18k8DmpmlkoTP-kl7svZ-bAL-MkRyvEQBoCQBMQAvD_BwE

Requirement through regulation

Laws are created by Congress to provide guidance and regulations that can help improve our lives in various ways. One example of this is the creation of the Pesticide Registration Improvement Act (PRIA) in 2003 and has since been renewed several times. The latest renewal, PRIA 4, was signed into law in March 2019 and is effective until 2023. [158]

PRIA is a significant piece of legislation that included a new section 33 under the Federal Insecticide, Fungicide, and Rodenticide Act (FIFRA). Section 33 of FIFRA established a registration fee for pesticide applications, which provided funding to support the registration and evaluation of new pesticides. This fee allowed the Environmental Protection Agency (EPA) to expedite the registration process for new pesticides, as well as fund research and development of safer and more effective pesticides.

In addition to PRIA, Congress also passed the 2004 Consolidated Appropriations Act. This act included funding for a variety of federal agencies, including the EPA. This funding allowed the EPA to continue its efforts to protect human health and the environment through the regulation of pesticides.

Together, these two acts provided important guidance and funding to support the regulation and evaluation of pesticides in the United States. By establishing registration fees for pesticide applications and providing funding for research and development, Congress has helped to ensure that the pesticides we use are safe and effective for both humans and the environment.

Before the implementation of section 33 by the EPA, approximately 200 companies manufactured pesticides in the United States. However, after the introduction of the new registration fee, only the largest and most financially stable companies were able

---

158 https://www.epa.gov/pria-fees/pria-overview-and-history

to afford to continue manufacturing pesticides. These companies included Syngenta Ag, UPL Corporation Ltd., Bayer Corp, Corteva, Inc., and Nufarm Ltd. [159]

This change had a significant impact on the pesticide industry, as it meant that smaller and newer businesses were at a disadvantage compared to larger, more established companies. The introduction of the registration fee also meant that there was a shift in the way pesticides were regulated in the US, with a greater emphasis on the role of the Environmental Protection Agency (EPA) in overseeing the registration process.

Despite the initial challenges faced by smaller companies, the registration fee has played an important role in helping to ensure that pesticides are properly regulated, and that only safe and effective products are made available to the public. The EPA continues to work closely with manufacturers to ensure that all pesticide products meet the necessary safety and efficacy standards, and that they are registered in a timely and efficient manner.

Laws are designed to improve the lives of American citizens and regulate business practices. In the United States, the vast majority of businesses are small and medium-sized enterprises (SMEs), with less than 500 employees. In fact, according to IBISWorld, 99.7% of all businesses fall into this category. [160]

This is significant data to consider when analyzing the impact of laws on companies. When a law is enacted, it may not affect smaller businesses with fewer than 500 employees, as they may not have the same level of resources and may not be subject to the same regulations. However, laws that impact larger corporations can still have a ripple effect on the broader economy, and SMEs may feel the impact indirectly.

---

159 https://www.epa.gov/pesticide-registration/pesticide-registration-manual-chapter-5-registration-fees
160 https://www.ibisworld.com/industry-statistics/number-of-businesses/pesticide-manufacturing-united-states

In the case of the Consolidated Appropriations Act [161] and the Pesticide Registration Improvement Act, [162] which introduced a registration fee for pesticide applications under the Federal Insecticide, Fungicide, and Rodenticide Act, [163] the impact on smaller businesses may have been minimal. As noted earlier, the five largest pesticide manufacturers in the US are Syngenta Ag, UPL Corporation Ltd., Bayer Corp, Corteva, Inc., and Nufarm Ltd. [164] These major corporations make up less than 0.03% of the labor force, and therefore the vast majority of the pesticide manufacturing industry consists of SMEs.

However, this does not mean that laws and regulations do not impact the chemical industry as a whole. According to the EPA, chemical industry shipments in the US are valued at over $555 billion per year. Any changes to laws and regulations in this sector can have a significant impact on the overall economy, even if they do not directly affect smaller businesses. [165]

The PRIA introduced a registration fee for pesticide applications, as noted earlier. However, this law also included a provision for waiver fees for companies producing pesticides with fewer than 500 employees, based on their gross chemical sales.

While laws may have a different impact on small and large businesses, it is important to consider the broader economic implications of any changes to regulations. The chemical industry is a major contributor to the US economy, and any changes to laws and regulations in this sector must be carefully considered.

---

161 https://www.congress.gov/bill/117th-congress/house-bill/2617
162 https://www.congress.gov/bill/116th-congress/senate-bill/483/text
163 https://www.epa.gov/laws-regulations/summary-federal-insecticide-fungi-cide-and-rodenticide-act
164 https://www.ibisworld.com/united-states/market-research-reports/pesticide-manu-facturing-industry/
165 https://www.ibisworld.com/industry-statistics/number-of-businesses/pesti-cide-manufacturing-united-states/

In the world of business, many companies have subsidiary companies that they own. For example, Company A may own Company B and Company C, even though they operate as separate entities. These corporate ownership structures can complicate the application of laws and regulations, as it may be unclear how to classify each company and apply any waivers or exemptions.

In the case of PRIA, the application of the waiver fees may be affected by these corporate ownership structures. When applying the rules, Congress intended to provide waivers to companies with fewer than 500 employees, but they did not account for the possibility that a larger company may have multiple subsidiary companies with fewer than 500 employees.

In this example provided, Company A, Company B, and Company C are all separate entities, even though they are owned by the same parent company. When applying the rules of PRIA, it may be unclear whether these companies should be considered as a single entity or separate entities for the purposes of the waiver fees.

As a result, the application of the law may be more complicated than Congress intended, and it may result in fewer companies receiving waivers than anticipated. In fact, according to some estimates, as much as 89% of the companies with fewer than 500 employees may not be eligible for the waivers, due to the complexities of corporate ownership structures.

This highlights the importance of considering the unintended consequences of laws and regulations, and the need for careful planning and consideration when drafting new legislation. While laws may be intended to provide benefits or protections, they may not always be applied as intended, and may have unintended consequences that need to be addressed.

PRIA provides a fee-for-service system that allows the Environmental Protection Agency (EPA) to collect fees from pesticide registrants to fund the registration process. PRIA establishes different fee categories for different types of applications, and also provides fee waivers and reductions for small businesses, federal and state agencies, and certain types of applications.

PRIA also includes provisions for funding farm worker protection activities, including surveys on farm worker employment, health, and living conditions, state oversight of farm worker illnesses and injuries, training programs and materials for farm workers to reduce risks of pesticide use, and training for health care providers to recognize and treat pesticide poisonings.

PRIA 2, which was enacted in 2007, expanded the number of fee categories and continued funding for farm worker protection activities. PRIA 3, which was enacted in 2012, expanded the number of fee categories again, established footnotes for covered PRIA categories, and continued funding for farm worker protection activities, partnership grants, and pesticide safety education programs.

PRIA 4, which is currently in effect, includes several updates to the registration process. These updates include the establishment of a new fee category for products with a reduced risk to human health or the environment, as well as changes to the fee categories for antimicrobial and public health pesticides. PRIA 4 also includes provisions for the electronic submission and review of labels and confidential statements of formula, and enhancements to the endangered species database. (See appendix for how the rules are applied.)

Grains feed to livestock

Corn and soybeans are two of the most widely used crops in the world, with corn accounting for about 40% of total domestic corn use in the United States, [166] while soybeans make up almost 80% of the world's soybean crop, much of which is used to feed livestock such as beef, pork, and chicken. [167] However, there are concerns about the expansion of soybean production, as millions of acres of land are being cleared to make way for soybean cultivation. This is resulting in the destruction of forests, prairies, and grasslands, which is endangering indigenous species.

Genetically engineered corn and soybean are two main components in livestock feed. 'A dairy livestock diet in the US usually contains 70% corn and 10% dehulled soybean meal, and poultry diets consist of 35% soybean meal and 65% corn grain (Van Eenennaam & Young, 2014). Poultry flocks, consuming about half of all soybean meal produced, are the single largest domestic consumer of soybean meal, followed by swine (Van Eenennaam & Young, 2014). The USDA (2018b) estimates that 87% of soybean and 57% of corn grain produced are used in livestock diets around the world each year.' [168]

Cows can be exposed to pesticides when they eat grass. This can be dangerous for the cows, as the pesticides can be toxic and can cause health problems. It is important to be aware of the potential risks associated with pesticide use and to take steps to minimize exposure to cows.

Pigs should not eat grains applied with pesticides. Pesticides can be toxic to animals and can cause health problems in animals and humans. It is best to feed pigs organic grains that have not been applied with pesticides.

---

166 https://www.ers.usda.gov/topics/crops/corn-and-other-feed-grains/feed-grains-sector-at-a-glance/
167 https://wwf.panda.org/discover/our_focus/food_practice/sustainable_production/soy/
168 https://www.ncbi.nlm.nih.gov/pmc/articles/PMC7386766/

In addition to land clearance, large amounts of pesticides are being applied to soybean crops to protect them from pests and diseases. When it rains, the chemicals run off into local tributaries, which can have a detrimental effect on other plants and animals in the surrounding areas. Furthermore, many of these pesticides are being used on other crops, such as wheat. In 2019, surveyed farmers used 87 different pesticides on winter wheat acres, 63 different ingredients on spring wheat acres, and 52 different pesticides on durum wheat acres. This raises concerns about the potential dangers of using so many different chemicals on crops. [169]

One of the issues with using so many different pesticides is that chemical companies submit each chemical to the Environmental Protection Agency (EPA) for approval one at a time. Therefore, the EPA grants approval to each pesticide and herbicide one chemical at a time, without considering the potential interactions between different chemicals. This is particularly concerning when so many different chemicals are being used on the same crop, as it is impossible to know how they will interact with one another. In fact, the number of potential chemical combinations created by using 87 different chemicals on one crop in one season is estimated to be 5.4723640075e+168 - a number so large that it is impossible to estimate or make an educated guess as to the potential dangers of these chemical combinations.

Given these concerns, there is a growing movement towards reducing the use of pesticides and promoting sustainable agricultural practices. This includes using crop rotation, natural pest control methods, and other strategies that can reduce the need for harmful chemicals. By promoting sustainable agriculture, we can help to protect our environment, our food supply, and our health.

---

169 https://www.nass.usda.gov/Surveys/Guide_to_NASS_Surveys/Chemical_Use/2019_Field_Crops/chem-highlights-wheat-2019.pdf

Here is the problem of using all these different pesticides on wheat crops. Chemical companies who produce these pesticides make submissions to the environmental protection agency (EPA) for approval one chemical at a time. EPA grant approval one of pesticides and herbicides one chemical at a time.

We need to start asking are all these different chemicals required to grow one crop of grain. Personally, I like to eat meat however, there are people who would like to ban feeding grains to livestock because about 37% of the world's grain and 66% of U.S. grain is used for livestock feed. Global livestock feed eats 86% of foods grains, like sugarcane tops we cannot digest as humans. The same study concluded that 46% of global livestock feed comes from grass and leaves. [170]

Two hundred million tons of pesticides are applied on grains fed to livestock in the United States in a report published in 2022. [171] The last report published in 2018 shows 235 million tons applied on crops fed to livestock. In October 2020 public-interest groups sued the U.S. Environmental Protection Agency (EPA) today over its decision to reapprove atrazine (is a chlorinated triazine systemic herbicide) and "It is fact and scientifically documented, that atrazine exposure "chemically castrates" frogs, impairs fish reproduction, and can result in birth defects and cancer in humans." [172] It is time the Unites States is working with the European Union to join together and places a ban the same pesticides. In November 2020 both public-interest groups and American Soybean Association and Plains Cotton Growers in Texas, filed lawsuits against the EPA about the use of pesticides on crops. Public-interest groups want the pesticide use to stop, and growers want to use pesticides without repercussions to make as much profit as possible.

---

170 https://www.sacredcow.info/blog/qz6pi6cvjowjhxsh4dqg1dogiznou6
171 https://biologicaldiversity.org/w/news/press-releases/new-report-more-than-200-million-pounds-of-pesticides-in-us-are-applied-to-crops-grown-to-feed-animals-on-factory-farms-2022-02-22/
172 https://beyondpesticides.org/dailynewsblog/2020/11/lawsuit-launched-against-epa-approval-of-toxic-herbicide-atrazine/

"Cotton and soybean growers planting in the nearly 300 counties nationwide potentially inhabited by listed species are subject to 310-foot downwind application buffers, and a 57-foot omnidirectional buffer, each ostensibly designed to achieve [Endangered Species Act] compliance," the ASA/Plains Cotton Growers lawsuit says. "Growers must also abide by a 240-foot, universally controlling, downwind FIFRA application buffer." [173]

It is important to present both sides of a discussion. However, if all these crops are being applied with pesticides and we as humans are eating these crops applied with unknown pesticides and combination of pesticides being applied on crops what is this doing to our health as adults and how does this effect children's health. This use of pesticides does not analyze the multiple combination of pesticides applied on the same crop and how this affects all living thing downstream from where pesticides are applied.

Growers of food for livestock states there is the following we cannot eat as humans. Protein companies argue livestock feed consists of 86% things we humans cannot eat and half comes from grass. [174]

LIVESTOCK TURN FOOD WE CANNOT EAT INTO PROTEIN

Grass and leaves feed to livestock	*46%*
Crop residuals	*19%*
Fodder crops	*8%*
Oil seed cakes	*5%*
By-products	*5%*
Other non-edible	*3%*
Grains	*13%*
Other edibles	*1%*

source: www.sacredcow.info

---

173 https://www.agri-pulse.com/articles/14813-epa-facing-lawsuits-over-atrazine-dicamba

174 https://www.sacredcow.info/blog/qz6pi6cvjowjhxsh4dqg1dogiznou6

Milk

In Vermont, dairy farms are required to keep records on spraying pesticides. From 1999 to 2012 Vermont's dairy farmers applied more than 2,533,329 pounds of metolachlor, atrazine, and simazine to their cornfields. Both atrazine and simazine are banned from use in the European union and metolachlor is under review. Syngenta Group renamed from "China Chemical (Shanghai) Agricultural Technology Corporation Ltd." to "Syngenta Group Co., Ltd. are the makers of atrazine. In 2012 Syngenta Group Co., Ltd were sued by twelve states for polluting drinking water and won. Vermont did not join this lawsuit and farmers in Vermont continued to use atrazine and even larger amounts of metolachlor. [175] We need to change our state's legislative agendas and have them on the same page to ban pesticides. Milk is considered one of the best nutrients for children and adults. Milk and dairy products (made of poor quality) or adulteration as there is a history in the world. New York 1850 tainted milk killed 8,000 infants and in China 2008 infant milk products contained melamine. [176]

As recent as 2016 milk adulteration has taken place in America when "U.S. Marshals Service seized more than 4 million pounds of [milk] product produced by Valley Milk Products LLC (Valley Milk) of Strasburg, Virginia." The FDA is using an investigative bacterial typing tool called whole genome sequencing (WGS) to link the samples collected in Valley Milk's facility over time. Implementing WGS technology can show the relationship of bacterial pathogens found in the environment, food source or people who became ill from eating contaminated food. [177]

---

175 https://regenerationvermont.org/vermonts-industrial-dairying-marketing-vs-reality/
176 https://foodsafetyandrisk.biomedcentral.com/articles/10.1186/s40550-016-0045-3
177 https://www.fda.gov/news-events/press-announcements/food-regula-tors-seize-adulterated-milk-products-food-safety-violations

In the 1930ies vitamin D was added to milk to prevent rickets among poor people. It is not mandatory in the United States, but most manufacturers of milk add vitamin D. By law in Canada and Sweden add vitamin D to cow's milk. [178]

The approval, which amends existing food additive regulations, will allow manufacturers to voluntarily add up to 84 IU/100g of vitamin D3 to milk, 84 IU/100g of vitamin D2 to plant-based beverages intended as milk alternatives, and 89 IU/100g of vitamin D2 to plant-based yogurt alternatives. [179]

In July 2016, FDA approved an increase to the amount of vitamin D that may be added as an optional ingredient to milk and approved the addition of vitamin D to beverages made from edible plants intended as milk alternatives, such as beverages made from soy, almond, and coconut, and edible plant-based yogurt alternatives. [180] Vitamin D was already authorized for use in soy beverages, but today's approval increases the authorized amount for such beverages that are intended as milk alternatives.

Today there are over 20 chemicals that can be found in milk. Some of these chemicals are environmental and others are added to the pasteurization process. Chemical contaminants in raw and pasteurized human milk study revealed there are 20 chemicals, including the persistent organic pollutants. [181]

These 20 compounds include:

Chloramphenicol	**Niflumic acid**
Florfenicol	**Phenylbutazone**
Pyrimethamine	**Triclosan**
Thiamphenicol	**Carbamazepine**
Diclofenac	**Clofibric acid**
Flunixin	**β-blockers**

---

178 https://www.healthline.com/nutrition/vitamin-d-milk#fortification-reasons
179 https://www.fda.gov/food/food-additives-petitions/vitamin-d-milk-and-milk-alternatives
180 https://www.fda.gov/food/cfsan-constituent-updates/fda-approves-increase-amount-vitamin-d-milk-and-milk-alternatives
181 https://www.worldofchemicals.com/150/chemistry-articles/milk-is-a-cocktail-of-nearly-20-chemical-compounds.html

Ibuprofen	**Metoprolol**
Ketoprofen	**Propranolol**
Naproxen	**17α-Ethinylestradiol**

Other ingredients that can be added to milk:

Lactose	**Potassium**	**Vitamin B1**	**Vitamin K**
Casein	**Sodium**	**Vitamin B2**	**Zinc**
Fat	**Magnesium**	**Vitamin B6**	**Copper**
Calcium	**Chloride**	**Vitamin B12**	**Iron**
Phosphorus	**Vitamin A**	**Vitamin D**	**Folate**

In a study in 2020 shows milk in the United States contains growth hormones, antibiotics, and low to elevated levels of pesticides not found in organic milk. [182]

Researchers from Emory University and aimed to compare the levels of pesticides, antibiotics, and hormones in milk produced through conventional and organic methods. The study analyzed a total of 160 milk samples collected from different regions of the United States.

The researchers found that organic milk had significantly lower levels of pesticides, antibiotics, and hormones compared to conventional milk. Specifically, the study found that conventional milk had a 50% higher concentration of antibiotics and a 60% higher concentration of hormones compared to organic milk. Additionally, the study found that conventional milk had a four times higher concentration of pesticides compared to organic milk. [183]

---

182 https://beyondpesticides.org/dailynewsblog/category/alternativesorganics/regenerative/
183 https://www.usatoday.com/story/money/2019/06/26/study-finds-organic-milk-cleaner-than-conventional-dairy/1482508001/

These findings are significant as they suggest that organic milk production methods may be a safer and healthier option for consumers. The study also highlights the importance of monitoring and regulating the use of pesticides, antibiotics, and hormones in conventional milk production to ensure the safety of milk products.

Adding data to support a theme of chemicals in our food system. 'Cow's milk is the most consumed product worldwide. However, due to various direct and indirect contamination sources, different chemicals and microbiological contaminants have been found in cow's milk.' [184]

The theme is pesticides in our food and cows eat the grass often applied with pesticides then we as consumers ingest these pesticides. In addition, mycotoxin (mycotoxins are toxic compounds that are naturally produced by certain types of fungi – fungi grow on a variety of different crops and foodstuffs including cereals, nuts, spices, dried fruits, apples, and coffee beans, often under warm and humid conditions) metals from the earth are being consumed by cows. [185] Then antibiotics and hormones are added at a feed yard next while processing milk to be pasteurized more metals and mycotoxin accumulate in the milk to be served on our tables.

The presence of chemicals in milk has been a cause for concern among health experts and consumers alike. Despite efforts to reduce chemical contamination in the milk supply, a study conducted in 2018 found that 19 out of 23 chemicals were still detected in pasteurized milk. These chemicals included both legacy chemicals like p,p'-dichlorodiphenyltrichloroethane (DDT) and emerging chemicals like phthalates. [186]

Brominated flame retardants, Polychlorinated biphenyls (PCBs), Organochlorine insecticides, Phthalate plasticizer, and Pyrethroid insecticides were among the five different chemical

---

184 https://www.ncbi.nlm.nih.gov/pmc/articles/PMC8822143/
185 https://www.ncbi.nlm.nih.gov/pmc/articles/PMC8228748/
186 https://pubmed.ncbi.nlm.nih.gov/29601252/

types detected in a gallon of milk. These chemicals are considered environmental contaminants and can have adverse health effects on humans and animals if ingested in large quantities. The study's findings highlight the need for continued efforts to reduce chemical contamination in milk and other food sources.

Raw cow's milk and goatmilk are available from a number of sources throughout the United States. Locations for finding real milk you can go to the following website realmilk.com. There are also other websites you can contact for fresh organic foods like localhens.com. Some other additives that can be included in dairy products.

Acesulfame potassium, also known as Ace-K, is an artificial sweetener that is commonly used in a variety of food and beverage products, as well as in other products like toothpaste and mouthwash. While it is approved for use by regulatory agencies such as the US Food and Drug Administration (FDA) and the European Food Safety Authority (EFSA), some studies have raised concerns about its potential health effects. [187]

For example, some animal studies have suggested that acesulfame (Ace-K) may be associated with an increased risk of tumors, including lung, breast, and organ tumors. In addition, some human studies have suggested that long-term consumption of Ace-K may be associated with side effects such as headaches, mental confusion, nausea, depression, and consequences on the liver and kidneys. [188]

However, it is important to note that the evidence on the potential health effects of acesulfame is not conclusive, and more research is needed to fully understand its impact on human health. Additionally, regulatory agencies such as the FDA and EFSA have determined that acesulfame is safe for human consumption at the levels typically found in food and beverage products.

---

187 https://www.medicalnewstoday.com/articles/318604#what-is-it
188 https://www.cancer.org/cancer/risk-prevention/chemicals/aspartame.html

Ultimately, whether or not to consume products containing acesulfame is a personal decision that should be made based on individual preferences and concerns. For those who are concerned about the potential health effects of artificial sweeteners, there are a variety of natural sweeteners available as alternatives, such as stevia, monk fruit extract, and erythritol. [189]

Annatto comes from the achiote tree (Bixa Orellana) is an orange-red food coloring or condiment made from the seeds of the achiote tree it is also has cytotoxic (toxic to living cells) properties such as wound healing, analgesic, hemostatic, and antioxidant activities. [190] [191]

Carmine (or E120) is extracted from the shells of the cochineal beetle and the shell contains carminic acid found in central and south America. Carmine is used as food coloring and can go from pink to red. [192]

Carrageenan's or carrageenin's are extracted from red edible seaweeds and is a hydrocolloid and can be used in dairy products as a thickening agent. Other types of thickening agents: alginate, pectin, gelatin, gellant, and agar. [193]

CITRIC ACID adjusts the pH using rennet as a complex set of enzymes that act on proteins in milk as it triggers coagulation. Traditionally used to separate milk into solid curds in cheese making. Fructose occurs naturally in milk and in fruits, vegetables, honey and can be used to sweeten foods like jellies, soft drinks, ice cream, candy, and other foods fructose, respectively.[194]

---

189 https://www.medicalnewstoday.com/articles/318604#which-foods-contain-it
190 https://pubmed.ncbi.nlm.nih.gov/34968663/
191 https://dontwastethecrumbs.com/30-additives-in-dairy-products-you-should-know-about/
192 https://www.imbarex.com/carmine-natural-coloring-in-drinks-and-dairy-products/
193 https://www.ncbi.nlm.nih.gov/pmc/articles/PMC3551143/
194 https://healthyeating.sfgate.com/lactose-fructose-intolerance-4457.html

Farm-to-fork

Our first farm-to-fork here in the United States was President Abraham Lincoln who created the U.S. Department of Agriculture (USDA) May 15, 1862, as he believed agriculture was critically important to domestic policy.

The need for a reboot of the USDA's farm-to-fork agency has been highlighted in a recent GAO report which found that the current health policy is disjointed across 21 different agencies, resulting in overlapping tasks and authority. [195] This lack of coordination has led to a poorly operated system, which is in dire need of reorganization. The Rockefeller Foundation, the American Action Forum, and the European Union have all recommended the need for a modernized food to fork system, which is transparent and efficient in its functioning.

Moreover, the lack of a modernized food to fork system can also affect our military readiness, making it a matter of national security. There are more than 20 agencies involved in supporting crops grown, which includes official offices related to farms and livestock. These agencies include the USDA, Environmental Protection Agency (EPA), National Institutes of Health (NIH), National Oceanic and Atmospheric Administration (NOAA), Centers for Disease Control and Prevention (CDC), and the Food and Drug Administration (FDA), among others. It is essential that these agencies work together in a coordinated manner to protect the food supply chain and ensure food safety for consumers.

The European union "farm-to-fork" approach control over key points in the food-producing chain. In the case of food-producing animals, besides nutritional value, feed must be free of contaminants that could be transferred through the food chain until they reach humans. Feed quality is directly influenced by environmental quality. Factors such as plant species, soil quality, fertilizing procedures, harvesting, processing, and storage. One farm in Baia Mare, Romania had the following metals lead, copper, cad-

---

195 https://www.agri-pulse.com/articles/18640-opinion-a-farm-to-fork-approach-to-fixing-fdas-food-program

mium, and zinc were in water, soil, feed, and milk samples collected from free-range cattle farms. [196]

In the US, there are regulations in place to ensure the safety of meat and dairy products for human consumption. However, the regulations for dairy products are not as strict as they are for meat. Milk cows are not typically tested for the presence of harmful chemicals or contaminants in the grass or feed they consume. Instead, the focus is on ensuring the cleanliness and safety of the milking process, as well as testing the milk itself for harmful bacteria or other contaminants. [197]

In contrast, regulations for meat production require more comprehensive testing and monitoring of the animals' diets and living conditions. The US Department of Agriculture (USDA) sets standards for the use of antibiotics and other drugs in livestock, as well as for the presence of pesticides, heavy metals, and other contaminants in animal feed.

However, there is still some concern among consumers and food safety advocates about the potential risks of consuming meat and dairy products from animals that have been exposed to pesticides, antibiotics, or other harmful substances. Some argue that more stringent regulations are needed to ensure the safety and quality of the food supply, particularly in light of emerging health and environmental concerns related to the use of these substances in agriculture.

Federal agencies that play a role in supporting food quality in the United States. For example, the National Institutes of Health (NIH) conducts research on the health effects of various food ingredients and contaminants, while the Department of Homeland Security (DHS) works to prevent intentional contamination of the food supply. The National Oceanic and Atmospheric Administration (NOAA) monitors seafood safety and quality, and the Federal Trade Commission (FTC) regulates food labeling and advertising. Furthermore, state and local agencies also play a role in ensuring food safety and quality, such as state health departments, state

196 https://www.ncbi.nlm.nih.gov/pmc/articles/PMC6862208/
197 https://www.ncbi.nlm.nih.gov/books/NBK209121/

agricultural departments, and local health departments. Overall, the collaboration between these various agencies is critical in ensuring the safety and quality of the U.S. food supply.

Multiple agencies that play a role in supporting food quality in the United States. The National Institutes of Health (NIH) is one such agency, which conducts research on the health effects of various food ingredients and contaminants. The Department of Homeland Security (DHS) is responsible for ensuring the security and safety of the food supply, including preventing intentional contamination of food.

The National Oceanic and Atmospheric Administration (NOAA) monitors seafood safety and quality, ensuring that seafood products are free from contaminants and safe for human consumption. The Federal Trade Commission (FTC) regulates food labeling and advertising to ensure that consumers receive accurate information about the products they are buying.

State and local agencies also play a significant role in ensuring food safety and quality. State health departments oversee food safety inspections in restaurants and other food service establishments, while state agricultural departments regulate the production and distribution of food products within their respective states. Local health departments also have a role in food safety inspections and enforcement.

U.S. Food and Drug Administration (FDA), support agencies Center for Biologics Evaluation and Research (CBER) Center for Food Safety and Applied Nutrition (CFSAN), Office of Food Policy and Response, Office of Minority Health and Health Equity, and National Center for Toxicological Research (NCTR).

Centers for Disease Control and Prevention (CDC), support agencies Center for Global Health, Center for State, Tribal, Local and Territorial Support, Agency for Toxic Substances and Disease Registry.

Environmental Protection Administration EPA), Office of Chemical Safety and Pollution Prevention (OCSPP), Office of Enforcement and Compliance Assurance (OECA)

US Department of Agriculture (USDA) support agencies Agricultural Research Service (ARS) Animal and Plant Health Inspection Service, Agricultural Marketing Service (AMS), Economic Research Service, Farm Service Agency, Food and Nutrition Service, Food Safety Inspection Service, Foreign Agricultural Service, Forest Service, FPAC Business Center, National Agricultural Static Service, National Institute of Food an Agricultural, National Resources Conservation Service, Risk Management Agency, Rural Development.

Overall, the collaboration between all of these agencies is critical in ensuring the safety and quality of the U.S. food supply. By working together, they can identify and mitigate potential risks, respond quickly to outbreaks and other food safety incidents, and provide consumers with the information they need to make informed decisions about their food choices.

Pesticide studies

The National Resources Defense Council (NRDC) has looked at four different studies about a type of pesticide called "neonics." These studies looked at how people might be accidentally exposed to neonics, like from using products that have the pesticide or from living near farms that use it. The studies found that there is a link between this type of pesticide and a higher chance of having problems with how your brain or body develops. This could cause things like problems with your memory or shaking fingers. The studies also found that neonics could cause problems with how your heart or brain forms before you are born. [198]

A recent study published in The International Journal of Biochemistry & Cell Biology found that exposure to neurotoxic pesticides in the environment can increase the risk of Parkinson's Disease (PD) by disrupting the gastrointestinal (GI) system. PD is

---

198 EPA-HQ-OPP-2020-0306-0071_attachment_1.pdf

a neurological disease that affects the brain and nervous system, causing problems with movement and other functions. The study shows that pesticides can harm the GI system, leading to an increased risk of PD. This means that exposure to pesticides can be harmful to our health and contribute to the development of serious diseases. It is important to limit our exposure to pesticides and find safer ways to grow and produce our food. [199]

Today glyphosate patent has expired and can be used as an active ingredient along with other pesticides ingredients. There is an issue with glyphosate because manufacturers of pesticides now can make blends or recommend blends to farmers and ranchers. The point here is each pesticide is submitted for approval one at a time not a blend.

Along with a relatively new group of pesticide neonicotinoid family since 1999 includes acetamiprid, clothianidin, dinotefuran, imidacloprid, nitenpyram, nithiazine, thiacloprid and thiamethoxam. The European Union and United Kingdom severely restricted the use of most neonicotinoids due to potential negative impact on bees and other pollinators. In 2020, all but one neonicotinoid was no longer be approved for use. However, in January 2022, emergency authorizations have been allowed, by the United Kingdom Government under an emergency application of the Cruiser SB pesticide, which contains thiamethoxam, a neonicotinoid. [200]

One question: Why apply chemicals known as pesticides that kill, diminish, or remove pollinating species from our world. With over 1,000 different pesticides, what is being done to protect living species that help build crop production. May 13, 2021, Johns Hopkins Bloomberg School of Public Health published a report "The finding that glyphosate appears to have an adverse effect on insects by interfering with their melanin production suggests the potential for a large-scale ecological impact, including impacts on

199 https://beyondpesticides.org/dailynewsblog/2022/05/neurotoxic-pesticides-disrupt-gut-function-linked-to-parkinsons-disease-development/
200 https://researchbriefings.files.parliament.uk/documents/CDP-2022-0024/CDP-2022-0024.pdf

human health". Insects use melanin for several purposes including cuticle sclerotization (part of *an insect's chitinous exoskeleton*) and color patterning, clot formation, organogenesis, and innate immunity. [201] Insects use melanin use as part of their immune defenses against bacteria and parasites; by reducing melanin allowing species to become infected by common pathogens. [202]

Environmental Working Group, April 7, 2022, Shopper's Guide to Pesticides in Produce™ analyzes test data from the Department of Agriculture and Food and Drug Administration, shows 50 percent of potatoes, spinach, lettuce, and eggplant had detectable levels of at least one of three bee-killing neonic insecticides banned in the European Union but still allowed for use on U.S. produce. [203] (Hi readers, maybe we can contact our legislative representative and demand we stop applying these pesticides.)

While the EPA in April 22, 2022 stated "[they are] confident that the fruits and vegetables our children are eating are safer than ever." EWG tests are "based on results of nearly 45,000 samples of produce tested by the Department of Agriculture and the Food and Drug Administration." [204] Above we have independent testing of crops that have pesticides in food being delivered to stores for us to consume. Second these pesticides being applied to crops are killing the bee population. I like to eat, and these pesticides should be banned, and we can help save the bees.

We need to define an allergic reaction to food like fruits, vegetables, wheat, or grains, which is normally initiated by an overreaction to your immune system or to a protein found in a given food. [205] More than 70 percent of non-organic fresh produce sold in the United States contains residues of potentially harmful pesticide. Over 50 percent of potatoes, spinach, lettuce, and eggplant had detectable levels of at least one of three bee-killing neonic insecticides

---

201 https://onlinelibrary.wiley.com/doi/full/10.1111/pcmr.12590
202 https://publichealth.jhu.edu/2021/ingredient-in-common-weed-killer-impairs-insect-immune-systems-study-suggests
203 https://www.ewg.org/foodnews/summary.php
204 https://www.ewg.org/foodnews/faq.php
205 https://acaai.org/allergies/allergic-conditions/food/

banned in the European Union but still allowed for use on United States crops produce. [206] Stronger rules to ban these pesticides are required here in the United States.

Using EWG.org food news summary shows through multiple tests of foods that our US food system has pesticides in our non-organic foods. [207]

Twelve fruits and vegetables with pesticides and some with 1.8 times of the allowed pesticides. Ninety percent of samples tested showed positive residues of two or more pesticides. Are you scared?

1. Strawberries	*7. Bell and hot peppers*
2. Spinach	*8. Cherries*
3. Kale, collard and mustard greens	*9. Peaches*
4. Nectarines	*10. Pears*
5. Apples	*11. Celery*
6. Grapes	*12. Tomatoes*

On the bright side seventy percent (70%) of these fruits and vegetable samples had no detectable pesticide residues.

1. Avocados	*8. Honeydew melon*
2. Sweet corn	*9. Kiwi*
3. Pineapple	*10. Cabbage*
4. Onions	*11. Mushrooms*
5. Papaya	*12. Cantaloupe*
6. Sweet peas (frozen)	*13. Mangoes*
7. Asparagus	*14. Watermelon*
	15. Sweet Potatoes

In this 2023 report from Environmental Working Group (EWG) have developed an interactive map showing animal species that were analyzed for levels of PFAS (forever chemicals) in their bodies. Researchers found chemicals in species such as scorpi-

---

206 https://www.ewg.org/foodnews/summary.php
207 https://www.ewg.org/foodnews/faq.php

ons, pandas, Siberian tigers, turtles, horses, dogs, plankton, sea lions, wild boar, otters, and oysters. Research on people started 60 years ago and only now are they beginning to understand the cause and effect of PFAS on humans. With over 12,000 different chemicals to make thousands of commercial products, they never break down and are linked to liver, kidney, fetal difficulties, and other health risks. [208]

A Georgia based Dalan Animal Health, Inc. in January 2023 received conditional approval from Department of Agriculture to use a vaccine on honeybees; a potential method to stop the decline of honeybees and is the first vaccine for insects. [209]

When first reading that a company is doing research to vaccinate honeybees should send alarm bells to everyone. A vaccine to protect bees from pesticides and other potential diseases. Jurassic Park and we are starting with insects instead of animals.

A large amount of evidence from scientific studies links the decline of honeybees to pesticide use and the environmental impact we are facing. These show there is a long-term effect and pose a risk to the world's ecosystems. [210]

> We are only now beginning to understand the cause and effect of PFAS on humans.

208 https://www.theguardian.com/environment/2023/feb/22/animal-toxic-pfas-contamination-study?CMP=Share_AndroidApp_Other
209 https://www.nytimes.com/2023/01/07/science/honeybee-vaccine.html
210 https://www.centerforfoodsafety.org/issues/304/pollinator-protection/bee-decline-and-pesticide-use-248

Bad pesticides

Polychlorinated biphenyls (PCBs) are a significant concern in the food system due to their persistence in the environment and potential health effects. PCBs were used in a wide range of industrial and commercial applications, including as insulating fluids for electrical transformers, hydraulic fluids, and lubricants. Because of their persistence, PCBs can be found in many different types of food, including fish, meat, dairy products, and vegetables. [211]

PCBs are known to cause a range of health effects, including developmental and neurological problems, reproductive disorders, and cancer. Due to their potential for harm, the United Nations Environment Program launched an initiative in 2001 to eliminate PCB products by 2028 in an environmentally sound manner. As a result of this initiative, the international total ban of PCBs came into effect on May 17, 2004. [212]

Despite the ban, PCBs continue to be detected in the food system, particularly in fish and seafood. PCBs can accumulate in the fatty tissue of fish, which can then be passed on to humans who consume them. The FDA and EPA have set guidelines and regulations to help reduce PCB exposure, including advising people to limit their consumption of certain types of fish and seafood, particularly those that are high in fat, and to avoid eating fish caught in contaminated waterways. [213]

Overall, continued monitoring and regulation of PCBs in the food system is necessary to ensure the safety and health of consumers. [214]

PCBs cause adverse human health effects by disrupting immune, reproductive, nervous, and endocrine systems. The International Agency for Research on Cancer (IARC) categorizes PCBs as group 1 human carcinogens. [215]

211 https://www.ncbi.nlm.nih.gov/pmc/articles/PMC9323099/
212 https://www.mdpi.com/1660-4601/19/21/13923
213 https://www.epa.gov/pcbs/learn-about-polychlorinated-biphenyls-pcbs
214 https://www.unep.org/explore-topics/chemicals-waste/what-we-do/persistent-organic-pollutants/pcbs-forgotten-legacy
215 https://www.ncbi.nlm.nih.gov/pmc/articles/PMC7759298/

International Agency for Research on Cancer (IARC) classification, Compounds or physical factors assessed are classified in four groups based on the existing scientific evidence for carcinogenicity. Group 1: "Carcinogenic to humans" There is enough evidence to conclude that it can cause cancer in humans. IARC definition and list of compounds. Group 2A: "Probably carcinogenic to humans" There is strong evidence that it can cause cancer in humans, but at present it is not conclusive.

IARC definition and list of compounds, Group 2B: "Possibly carcinogenic to humans" There is some evidence that it can cause cancer in humans but at present it is far from conclusive. [216]

IARC definition and list of compounds would include infection conditions, chemical substances, radiation and physical agents, complex mixtures or agents, exposure, [217] Carcinogenic risk' in the IARC Monographs series is taken to mean the probability that exposure to an agent will lead to cancer in humans.

Glyphosate

Glyphosate is no longer under patent with the Monsanto and Bayer company. Forty-one countries have banned or are banning the use or importation of glyphosate. Notability missing is the United States from banning glyphosate use. In Sri Lanka, India, and Central America glyphosate has become suspect cause of widespread kidney disease. A WARNING label is printed this herbicide as it is considered moderately toxic as a requirement.

Perhaps you are already familiar with this particular organophosphate, as it is the active ingredient in Roundup, Monsanto's popular pesticide, which has been making legal/ medical news in recent years. Glyphosate is no longer under patent with the

216 https://www.greenfacts.org/glossary/ghi/iarc-classification.htm
217 https://en.wikipedia.org/wiki/IARC_group_1

company, and now it is used as an active ingredient in many similar products available today. Austria France and Germany have banned Glyphosate. [218]

Studies from over two decades ago determined that glyphosate is relatively non-toxic to most plants and animals. However, the herbicide is considered moderately toxic and requires the signal word WARNING to be printed on the label due to the significant irritation it can cause in the eyes.

But these studies (which made observations not exceeding two years in time) did not consider the long-term effects, which we have since seen. Over time, through constant exposure, we have now seen many hazards present themselves.

However, it is still commonly used in the United States.

Additionally, glyphosate is commonly combined with other toxic herbicides such as atrazine and neonicotinoids. This is one reason why most of our food supply is contaminated with multiple pesticides when glyphosate is found.

And it is these chemical combinations that are dangerous and highly hazardous to all life on the planet. According to the USDA, more than 225 different pesticides have been found on many commonly consumed fruits, vegetables, and grains in the United States. [219]

Glyphosate itself is found in trace amounts in hundreds of the most common foods, including popular brand name granola bars, hummus, orange juice, and more. But even trace amounts add up, and when done multiple times daily, the effects start to take hold that much quicker.

Glyphosate is a widely used herbicide and the active ingredient in products such as Roundup. In 2015, the International Agency for Research on Cancer (IARC), a specialized agency of the World Health Organization, classified glyphosate as "probably carcinogenic to humans" (Group 2A). This classification was

---

218 https://www.ncbi.nlm.nih.gov/books/NBK470430/
219 https://impactful.ninja/worst-pesticides-for-the-environment/

based on limited evidence of cancer in humans from real-world exposures and sufficient evidence of cancer in experimental animals from studies of pure glyphosate.

In addition to cancer, IARC also found strong evidence for genotoxicity, or damage to genetic material, for both pure glyphosate and glyphosate formulations. IARC's evaluation is based on the systematic assembly and review of all publicly available and pertinent studies by independent experts who are free from vested interests. Their evaluation follows strict scientific criteria, and the classification system is recognized and used as a reference all around the world. [220]

To reach these conclusions, IARC reviewed about 1000 studies, including studies on people exposed through their jobs, such as farmers, and experimental studies on cancer and cancer-related effects in experimental systems. The IARC classification has been a subject of debate and controversy, with some studies suggesting that glyphosate is not carcinogenic, while others support IARC's findings. Nonetheless, the classification has led to regulatory actions in some countries, such as restrictions on the use of glyphosate in agriculture and public parks. [221]

Pesticide spraying near schools have been known to cause acute health effects include stinging eyes, rashes, blisters, blindness, nausea, dizziness, diarrhea, and death. Examples of known chronic effects are cancers, birth defects, reproductive harm, immunotoxicity, neurological developmental toxicity, disruption of the endocrine system and even death in extreme cases. [222]

Glyphosate molecules have been used by farmers since 1974 and genetically modified organisms (GMOs) have been designed to allow this glyphosate molecule to be applied on food products to help the food supply. Since its introduction its health impacts have been heavily debated.

220 https://pubmed.ncbi.nlm.nih.gov/27677670/
221 https://www.iarc.who.int/featured-news/media-centre-iarc-news-glyphosate/
222 https://www.pesticidereform.org/

The function on how glyphosate works by inhibiting an enzyme found in plants, leading to their death. However, it is also known to kill non-targeted plants and is, therefore, used with caution.

Many companies engineer crops and seeds that are resistant to glyphosate to facilitate its broader usage. GMO crops are prohibited in Europe as part of its non-GMOs policy. [223] In the Unites States we continue to use GMO glyphosate resistant plants and continue to spray them with glyphosate.

Studies have shown glyphosate degrades at a relatively rapid rate in most soils, with half-life estimated between 7 and 60 days. However, extensive usage of glyphosate may pose chronic and remote hazards to the ecological environment. Glyphosate was detected in 36% of a total of 154 water samples collected from Midwestern U.S. states, where glyphosate is extensively used on corn. [224]

The endocrine system is a messenger system comprising feedback loops of the hormones released by internal glands of an organism directly into the circulatory system, regulating distant target organs. In vertebrates, the hypothalamus is the neural control center for all endocrine systems. In humans, the major endocrine glands are the thyroid gland and the adrenal glands. The study of the endocrine system and its disorders is known as endocrinology. [225]

The issue of whether glyphosate can act as an endocrine disruptor has been the subject of much debate and controversy. Endocrine disruptors are chemicals that can interfere with the body's hormonal system, potentially leading to a range of adverse health effects, such as developmental abnormalities, reproductive problems, and cancer.

---

223 https://www.alcimed.com/en/alcim-articles/end-glyphosate-europe-perspective-alternatives/
224 https://www.ncbi.nlm.nih.gov/pmc/articles/PMC6918143/
225 https://en.wikipedia.org/wiki/Endocrine_system

In the case of glyphosate, the Endocrine Disruptor Screening Program (EDSP) of the United States Environmental Protection Agency (EPA) and the European Food Safety Authority (EFSA) have conducted extensive evaluations of the available evidence to determine whether glyphosate has endocrine disrupting properties. [226] According to the EDSP and EFSA reports, there is currently no sufficient evidence to support the claim that glyphosate acts as an endocrine disruptor. However, some studies have suggested that glyphosate can interfere with certain hormonal pathways in the body, particularly those involving the thyroid gland and the sex hormones.

Given the complexity of the endocrine system and the potential for long-term effects, the debate over the endocrine disrupting effects of glyphosate is likely to continue. However, at present, the regulatory agencies responsible for assessing the safety of pesticides have not found sufficient evidence to support the claim that glyphosate is an endocrine disruptor. [227]

Pesticides can cause short-term adverse health effects, called acute effects, as well as chronic adverse effects that can occur months or years after exposure. Examples of acute health effects include stinging eyes, rashes, blisters, blindness, nausea, dizziness, diarrhea, and death. Examples of known chronic effects are cancers, birth defects, reproductive harm, immunotoxicity, neurological and developmental toxicity, and disruption of the endocrine system.

In the article Controversies on Endocrine and Reproductive Effects of Glyphosate and Glyphosate-Based Herbicides discusses the potential effects of glyphosate exposure on reproductive health, both in animal models and in humans. The authors suggest that data on reproductive outcomes, combined with evidence of glyphosate's effects on hormone production, support the possi

---

226 https://www.epa.gov/endocrine-disruption/endocrine-disruptor-screening-program-edsp-policies-and-procedures
227 https://www.ncbi.nlm.nih.gov/pmc/articles/PMC8006305

bility of a causal link between glyphosate exposure and pregnancy-related problems, such as miscarriage and shortened pregnancy duration. [228]

The authors note that animal studies have shown disrupted uterine differentiation and increased miscarriage rates after exposure to glyphosate-based herbicides (GBHs). They also cite studies that have found associations between glyphosate exposure and shortened pregnancy duration in humans. Suggesting that the endocrine-disrupting effects of glyphosate may play a role in these reproductive effects, as glyphosate has been shown to disrupt steroid hormone production. They argue that these effects are biologically plausible and should be further studied in order to better understand the potential risks of glyphosate exposure to reproductive health. [229] Then Herbicides can act by inhibiting cell division, photosynthesis, or amino acid production or by mimicking natural plant growth hormones, causing deformities (Ross and Childs 1996). Application methods include spraying onto foliage, applying to soils and applying directly to aquatic systems.[230]

A systematic review and meta-analysis of studies examining the association between glyphosate exposure and Non-Hodgkin Lymphoma (NHL). The review found that exposure to glyphosate was associated with a 41% increased risk of NHL, particularly among people with weakened immune systems. The study included a total of 15 studies, with a combined sample size of over 300,000 participants. However, it's important to note that the abstract only provides a brief summary of the study's findings, and the full paper should be read to fully understand the study's methods and conclusions. Additionally, correlation does not necessarily imply causation, so further research is needed to determine the exact nature of the relationship between glyphosate and NHL. [231]

228 https://www.ncbi.nlm.nih.gov/pmc/articles/PMC5844093/
229 https://www.frontiersin.org/articles/10.3389/fendo.2021.627210/full
230 https://www.epa.gov/caddis-vol2/herbicides
231 https://pubmed.ncbi.nlm.nih.gov/31342895/

The debate surrounding glyphosate intensified in 2015 when the International Agency for Research on Cancer (IARC), a branch of the World Health Organization, classified glyphosate as "probably carcinogenic to humans." The classification was based on limited evidence of cancer in humans and sufficient evidence of cancer in experimental animals. [232]

The classification sparked a wave of regulatory and legal action against glyphosate use. The Environmental Protection Agency (EPA), which is responsible for regulating pesticides in the United States, has maintained that glyphosate is safe for use based on current uses and that there is no evidence that it causes cancer. However, several states and countries have banned or restricted its use.

Bayer, the pharmaceutical company that acquired Monsanto, the original manufacturer of glyphosate-based herbicides, maintains that glyphosate is safe and that it has been thoroughly tested and approved by regulatory agencies around the world. However, the company has faced numerous lawsuits filed by people who claim that they developed cancer after being exposed to glyphosate.

Despite the ongoing debate, glyphosate remains one of the most widely used herbicides in the world, and its use is expected to continue for the foreseeable future. The controversy surrounding its safety and potential health impacts highlights the need for continued research and regulatory oversight to ensure the safety of agricultural chemicals. [233]

Recent studies have questioned glyphosate safety and international agencies have conflicting opinions about its effects on human health, mainly as an endocrine-disrupting chemical (EDC) and its carcinogenic capacity. [234]

Plant growth regulators (PGRs) are chemical compounds that are used to modify the growth, development, and physiologi

---

232 https://www.iarc.who.int/iarc-monograph-on-glyphosate-other-related-information/
233 https://www.nbcnews.com/data-graphics/toxic-herbicides-map-showing-high-use-state-rcna50052
234 https://pubmed.ncbi.nlm.nih.gov/33131751/

cal responses of plants. However, the use of PGRs in agriculture has been associated with several side effects that are harmful to human health. PGR residues in agricultural products have been found to cause various health problems such as hepatotoxicity, nephrotoxicity, genotoxicity, neurotoxicity, carcinogenicity, and teratogenicity. [235]

Hepatotoxicity refers to liver damage caused by exposure to certain chemicals, including PGRs. [236] Nephrotoxicity is the rapid deterioration in kidney function, which can lead to kidney damage and failure. [237] Genotoxicity occurs when chemicals damage the genetic information within a cell, causing mutations that may lead to cancer. [238] Neurotoxicity can alter the normal activity of the nervous system, leading to a variety of symptoms such as numbness, tingling, weakness, and even paralysis.

Moreover, PGRs are suspected of disrupting the function of human and animal reproductive systems. The potential adverse effects of PGRs on human health have led to increasing concern among consumers and regulatory authorities, and efforts are being made to minimize the use of these chemicals in agriculture. [239]

The co-occurrence of pesticide residues in our food system is a growing concern for public health. Pesticides are used to control pests and increase crop yields in agriculture, but their residues can persist on crops and in the environment for extended periods. Exposure to these residues can occur through various routes, including ingestion of contaminated food and water, inhalation, and dermal contact. National Library of Medicine 2021 [240]

The presence of multiple pesticides in our food and environment raises concerns about the potential health effects of their mixture. Studies have shown that exposure to mixtures of pesticides can lead to a greater toxicological impact than exposure to

---

235 https://www.ncbi.nlm.nih.gov/pmc/articles/PMC7544804/
236 https://pubmed.ncbi.nlm.nih.gov/33426623/
237 https://www.ncbi.nlm.nih.gov/pmc/articles/PMC6598532/
238 https://pubmed.ncbi.nlm.nih.gov/12787816/
239 https://pubmed.ncbi.nlm.nih.gov/30168961/
240 https://www.ncbi.nlm.nih.gov/pmc/articles/PMC7904562/

individual pesticides alone. The toxic effects of pesticide mixtures can be additive, synergistic, or antagonistic, depending on the specific combination of pesticides and the doses involved.

However, the assessment of the health effects of pesticide mixtures is complicated and challenging, as the interactions between different chemicals can be complex and nonlinear. The mathematical models used to predict the effects of pesticide mixtures on human health are often too large and complex to provide definitive answers.

To ensure the safety of pesticides, companies are required to submit proprietary reports to the Environmental Protection Agency (EPA). These reports include information on the toxicity of individual pesticides and the potential risks of exposure to humans and the environment.

Previously, companies would fill out Form U, which contained a series of questions about the pesticide being registered. The EPA has now implemented a new electronic system to expedite the registration of new pesticides. The questions asked on Form U and the information required by the EPA vary depending on the type of pesticide and its proposed use. However, the ultimate goal is to ensure that the pesticide is safe for humans and the environment when used according to the label instructions.

Many studies submitted to regulatory agencies as proprietary reports, which often are not readily available to the public. Nonclinical (animal) toxicology and safety data can also be made available to the scientific public through the publication of results in the open scientific literature.[241] I am sure all of us have time to go to the Internet or library to research each of these proprietary documents submitted to government agencies.

---

241 https://www.iadclaw.org/assets/1/7/19.2_-_Exponent-_REVIEWED-_Glyphosate_commentary_-_JPHE-17-107.pdf

Toxicity

HOW PESTICIDES AND HERBICIDES EFFECT OUR HEALTH FROM FOOD CONSUMPTION.

The use of pesticides and herbicides in agriculture has become increasingly widespread over the years, and the potential health impacts of exposure through food consumption are a growing concern. Pesticides and herbicides are designed to be toxic to pests and weeds, but they can also have harmful effects on human health.

Pesticides are classified into four categories based on their toxicity levels. Category I pesticides are highly toxic and can cause severe irritation, while category II pesticides are moderately toxic and can cause moderate irritation. Category III pesticides are slightly toxic and only cause mild irritation, while category IV pesticides are practically non-toxic and do not cause any irritation. [242] The health impacts of pesticide exposure can vary depending on the type and amount of pesticide consumed, as well as the duration of exposure. Some of the potential health impacts of pesticide exposure include damage to the nervous system, reproductive problems, cancer, and birth defects.

Furthermore, exposure to herbicides like glyphosate has also been linked to adverse health effects, including an increased risk of cancer and reproductive problems. It is important for individuals to be aware of the potential risks associated with pesticide and herbicide exposure and take steps to minimize their exposure, such as washing produce thoroughly and choosing organic options when possible.

---

242 https://pubmed.ncbi.nlm.nih.gov/6814873/

Glyphosate is still a widely used herbicide and the active ingredient in several popular weed killers. The use of glyphosate has been heavily debated in recent years due to concerns over its potential health and environmental impacts. Glyphosate is available in several formulations, including as an acid, diammonium, isopropylamine (IPA), monoammonium, sodium, and trimesium.

Studies continue to suggest that exposure to glyphosate may be linked to several health issues, including cancer, reproductive problems, and neurotoxicity. In addition to the potential health risks, glyphosate has also been found to be hazardous to the environment, particularly to aquatic life and soil health.

The use of glyphosate has been controversial, with some arguing that it is a safe and effective herbicide, while others call for stricter regulation or a ban on its use. The debate around glyphosate and its impact on human health and the environment continues, and research in this area is ongoing.

The addition of these nine pesticides Cyproconazole, Propiconazole, Tioxafen, Propineb, Flupyradifurone, Noviflumuron, Calcium cyanide, Sodium cyanide, and Hydrogen cyanide since March 2018 has raised concerns about their potential impact on human health and the environment. [243] Cyproconazole and propiconazole, previously considered safe, are now considered presumed human reproductive toxicants. Tioxafen and propineb, previously considered possible carcinogens, are now considered probable carcinogens. Flupyradifurone is highly toxic to honeybees, and Noviflumuron is now considered a probable carcinogen. Calcium cyanide, sodium cyanide, and hydrogen cyanide have all been reclassified based on their toxicity levels, with hydrogen cyanide now classified as "fatal if inhaled." The addition of these pesticides to the list of potentially harmful chemicals highlights the

---

243 http://npic.orst.edu/chemicals_evaluated.pdf

need for continued research into their effects on human health and the environment. It also underscores the importance of regulation and monitoring of pesticide use to protect public health and the environment. [244]

Methomyl is a neurotic insecticide, meaning it can have a poisonous effect on nerve cells or tissue. Its carbamate class is very much like organophosphates. The EPA recently determined methomyl is presumed to be hazardous to more than 1,100 endangered species, including San Joaquin kit foxes, whooping cranes, and all protected species of salmon. [245]

Studies from Cornell and company have shown methomyl to be highly toxic to birds, bees, fish, and other aquatic invertebrates. It can also contaminate groundwater, and plants will take up methomyl through its roots and leave traces in food. [246]

The worst pesticides include Atrazine, Flupyradifurone, Hexachlorobenzene, Glyphosate, Methomyl, and Rotenone. [247]

> Exposure to glyphosate is linked to several health issues, including cancer, reproductive problems, and neurotoxicity.

244 https://pan-international.org/wp-content/uploads/PAN_HHP_List.pdf
245 https://www.epa.gov/ingredients-used-pesticide-products/methomyl
246 https://impactful.ninja/worst-pesticides-for-the-environment/
247 https://impactful.ninja/worst-pesticides-for-the-environment/

Pesticides used on plants

Atrazine is an herbicide that has been in use for more than 60 years, primarily to control weeds in corn and sorghum crops. However, its use has been controversial due to concerns over its potential impact on human health and the environment. Atrazine has been linked to a range of health issues, including birth defects, low sperm counts, fertility problems, and endocrine disruption. For example, water that has been contaminated will harm fish and amphibians by compromising their growth, weight, immune function, and behavior. [248]

Studies have shown that atrazine can interfere with the hormonal system, affecting the production of estrogen and testosterone. This can lead to a range of reproductive problems, including reduced fertility and increased risk of birth defects. Atrazine has also been linked to an increased risk of certain types of cancer, including prostate cancer and non-Hodgkin lymphoma.

Despite these concerns, atrazine continues to be used widely in the United States, with millions of pounds of the chemical being applied each year. In contrast, 44 countries have banned or restricted the use of atrazine due to concerns over its health and environmental impacts.

Proponents of atrazine argue that it is a safe and effective herbicide that plays an important role in agriculture. However, many environmental and health advocates continue to call for stricter regulations on the use of atrazine, or for its complete ban.

While this was a popular herbicide worldwide for quite some time, the EU banned atrazine in 2004. Later, the EPA "re-evaluated" it and approved it for use in 2009, noting its ability to 'break down quickly' in soil. However, PAN analysis using data from the United States Department of Agriculture determined that atrazine can be found in almost 90% of America's drinking water; and such figures imply that there is not much 'break-down' happening at all. [249]

---

248 https://www.ncbi.nlm.nih.gov/pmc/articles/PMC4137807/
249 https://www.panna.org/resources/atrazine-size-under-siege/

Perhaps the most significant proof of its detriment to the environment, studies have found that atrazine laced water was responsible for 10% of male frogs exposed turning female. And that particular study had used controlled contamination of 2.5 ppb (parts per billion), which is less than the EPA limit for drinking water in the United States. [250]

Flupyradifurone

The US EPA registered flupyradifurone as a new insecticide in 2015 with claims that it is "safer for bees" than organophosphates, neonicotinoids, pyrethroids, and other previously established insecticides. [251]

However, some noticed that flupyradifurone is chemically similar to neonicotinoids. Many suspected it possessed many of the same unwanted attributes as neonicotinoids and the other conventional choices, despite the safety claims. And a 2019 study of lethal and sublethal synergistic effects of flupyradifurone on honeybees determined just that.

HEXACHLOROBENZENE

This organochlorine fungicide was widely popular for many decades, used mostly as a protectant for grain, wheat, and other field crop seeds. Hexachlorobenzene (HBC) has not been produced in the United States as a commercial product for quite some time; however, since the 1970s, most HBCs are formed as a byproduct of producing chlorinated solvents and compounds and other pesticides. [252] HBCs have an incredibly high resistance to breakdown—both chemical and biological—and have contaminated

---

250 https://news.berkeley.edu/2010/03/01/frogs/
251 https://impactful.ninja/worst-pesticides-for-the-environment/
252 https://www.ncbi.nlm.nih.gov/pmc/articles/PMC5464684/

all parts of our environments. HBCs have been detected in the air, water, and soil. Plants, aquatic life, mammals, and people have tested positive for HBC contamination, with humans being exposed to it mostly through the food chain.

The HBC Biomonitoring Summary from the Centers for Disease Control and Prevention (CDC) cites specific data from many government agencies and environmental organizations considered when classifying HBC toxicity. But don't let all those acronyms, dates, and data detract you from what they're trying to pass off as anecdotal; the fact is that HBCs are harmful.

ROTENONE

Think going organic will eliminate hazards? Think again. Despite advocates for the industry claiming it has been taken off the market, this highly toxic chemical is used by organic farmers (most often unknowingly) as it is combined with pyrethrins (are pesticides found naturally in some chrysanthemum flowers) products. Although it is a natural chemical, it has high toxicity that comes with nasty consequences to both you and the environment.

Even PAN chose to overlook this deadly chemical that enhances the onset of Parkinson's Disease and kills bees. (But as we know, these organizations aren't looking to eliminate the problem, just documenting, and controlling it.) Also known to be related to causing Parkinson. [253]

A tiny amount of rotenone derived from lonchocarpus species and can kill every fish in your pond, yet it continues to be approved for commercial use. Worse yet, organic food does not get tested for (natural) chemical residue, so health risks cannot be properly assessed. [254]

---

253 https://www.nih.gov/news-events/news-releases/nih-study-finds-two-pesticides-associated-parkinsons-disease
254 https://impactful.ninja/worst-pesticides-for-the-environment

Top 10 classic herbicides

The following pesticides recently have been evaluated to have an effect on the organs of the human body. Studies of these pesticides and herbicides reveal their toxic effect on humans. Bottom line, more studies are required to make a definitive answer. However, with out individuals writing letters to their government representatives state and federal little will change.

Glyphosate revolutionized weed control when this nonselective herbicide was teamed up with the Roundup Ready trait.

Atrazine is a part of most weed control programs in corn today. It is inexpensive, used in new (and older) herbicide tank mix combinations, and supports conservation tillage. A little over 60 million pounds of atrazine was applied on corn and soybeans in the United States in 2018. [255]

2,4-D is more than 60 years old and still controls most major weed problems in corn. It also is versatile for use in soybeans as an early preplant application.

Simazine an herbicide that controls a wide range of grasses and broadleaf weeds. [256]

Metolachlor is a selective and systemic herbicide which controls weeds by inhibiting the synthesis of long chain fatty acids. [257]

Dicamba was introduced in the late 1960s and remains a versatile corn herbicide. [258]

Prowl (pendimethalin) continues to be a good product for control of annual grasses and small-seeded annual broadleaf weeds in corn and soybeans. [259]

Liberty is an alternative for glyphosate in LibertyLink crops and to date has limited weed-resistance issues. Recently, it has been called Ignite. [260]

---

255 https://thehill.com/changing-america/sustainability/environment/597140-factory-farms-use-a-quarter-billion-pounds-of/
256 https://www.cdpr.ca.gov/docs/risk/rcd/simazine_risks_exposure_groundwater.pdf
257 https://www.ncbi.nlm.nih.gov/pmc/articles/PMC4578801/
258 https://www.ncbi.nlm.nih.gov/pmc/articles/PMC7660157/
259 https://www.ncbi.nlm.nih.gov/pmc/articles/PMC2674312/
260 https://www.ncbi.nlm.nih.gov/pmc/articles/PMC6360103/

Pursuit is one of the first herbicides growers relied on for total post weed control programs. It garnered a significant market share in soybeans. [261]

Basagran was one of the first post-planting selective herbicides for soybeans. It provided control of many difficult annual broadleaf weeds.

Dual and Lasso were mainstays in the mid to late 1970s. They allowed farmers to apply residual herbicides without significant tillage, which gave rise to the conservation tillage movement. These and other similar products still provide excellent control of annual grasses and small-seeded broadleaf weeds in corn and soybeans.

Treflan was one of the first DNA herbicides commercially available. It was a dominant soybean product for many years and gave rise to the commercial development of several other DNA herbicides, including Tolban, Cobex, Sonalan, Prowl, Basalin, and Endurance. [262]

European Food Safety Authority (EFSA) is very similar to EPA, USDA and FDA in the US. Concentrates in areas of generic risk assessments and the assessment of applications for the authorization of regulated products, in particular in the areas of Animal Health and Animal Welfare, Biological hazards and Chemical contaminants, Pesticides, Plant health, Genetically Modified Organisms for food and feed uses, Food Additives, Food Contact Materials, Food Enzymes, Food Flavorings, Feed additives, Novel Foods, Nutrition and activities in the area of Social Sciences. scientists to assist its units in carrying out the preparatory work for scientific outputs in the areas of Animal Health and Animal Welfare, Biological hazards and Chemical contaminants, Pesticides, Plant health, Genetically Modified Organisms, Food Additives, Food Contact Materials, Food Enzymes, Feed additives, Novel Foods and Nutrition. [263]

261 https://www.farmprogress.com/crop-protection/top-10-classic-herbicides
262 https://www.farmprogress.com/herbicides/top-10-classic-herbicides
263 https://www.efsa.europa.eu/en/procurement/scientifictechnicalsupport

Clinical toxicology is a dynamic field of medicine; new treatment methods are developed regularly, and the effectiveness of old as well as new techniques is subject to constant review. Prevention of pesticide poisoning remains a much surer path to safety and health than reliance on treatment.[264]

Federal regulators estimate that glyphosate is likely to harm or kill 93 percent of the animals and plants protected under the Endangered Species Act, according to the report's findings. The second most used herbicide per the study was atrazine. A little over 60 million pounds of atrazine were placed on corn and soybeans in the United States in 2018, a 17 percent increase from the amount used in 2012, report findings show.[265]

Solutions

Farmers and ranchers who work with nature have recognized the importance of protecting the environment, not only for their livelihood but also for future generations. To achieve this, they invest time and money in controlling erosion, planting trees, and protecting streams and rivers. Farmers are also turning to integrated pest management (IPM) techniques, which rely on a combination of methods to control pests while minimizing the use of synthetic pesticides. These methods can include biological controls such as the use of beneficial insects, cultural controls like crop rotation and sanitation, and chemical controls such as pheromone traps and selective pesticides. The goal of IPM is to reduce pest populations to a level where they are no longer causing economic damage, while also minimizing the impact on the environment and human health. [266]

---

264 https://extension.missouri.edu/publications/g1915
265 https://thehill.com/changing-america/sustainability/environment/597140-factory-farms-use-a-quarter-billion-pounds-of/
266 https://www.usda.gov/oce/pest/integrated-pest-management

IPM strategies that aim to reduce pesticide use and promote ecological balance. This can involve techniques such as crop rotation, planting pest-resistant crops, using natural predators to control pest populations, and minimizing disturbance to natural habitats surrounding the farmland. Farmers should also prioritize the use of non-toxic and non-persistent pesticides, and always follow proper application guidelines to prevent overspray or drift onto non-target areas.

While pesticides are often seen as a quick fix to pest problems, there are more sustainable and environmentally friendly ways to control pests. In addition to regulating the production and use of pesticides, there is also a need for greater public education on the dangers of pesticides and the benefits of using alternative methods of pest control. IPM is an effective alternative that can help reduce the use of pesticides while still protecting crops from pests. IPM involves the using a combination of natural predators, crop rotation, cultural, biological, chemical control, and other methods that do not rely on synthetic chemicals.

The practice of Integrated Pest Management (IPM) is a key strategy used by these farmers to identify when pest control is required through monitoring. IPM recognizes that not all insects, weeds, and other living organisms require control and that there are hundreds of organisms that are beneficial to the land. It seeks to protect multiple species before an application of a pesticide is made. By managing the land before a threat occurs, through rotation of crops and choosing pest-resistant crops, farmers can increase profits and eliminate the need for costly pesticide applications. In addition, crop rotation can also help maintain soil health, improve crop yields, and reduce soil erosion. Overall, working with nature can help farmers and ranchers achieve sustainable production, protect the environment, and ensure that their land remains productive for generations to come. [267]

---

267 https://www.epa.gov/safepestcontrol/integrated-pest-management-ipm-principles

Developing method is an eco-management plan (EMP) this requires careful consideration of various factors. The first step is to identify the social values and priorities that should be considered. Second, protecting natural resources, preserving cultural heritage, and promoting economic development. Once these priorities have been established, it is important to create clear boundaries and spaces for the project to ensure that it stays within its intended scope.

The third step involves identifying the social benefits that will be achieved by the project. These could include job creation, improving the quality of life for local communities, and promoting sustainable development. The fourth step involves identifying and addressing any potential strains or tensions that may arise during the project. This could include conflicts over resource use, differing opinions on how to achieve goals, or concerns about the impact of the project on the environment.

The fifth step is to consider the biological results that can be achieved through the project. This may involve identifying and protecting endangered species, improving habitat quality, or reducing pollution levels. The sixth step is to assess the benefits and costs of the project and determine if it is sustainable over the long term. This could include evaluating the economic, social, and environmental impacts of the project and identifying any potential trade-offs or unintended consequences.

Finally, the seventh step involves gathering all the relevant scientific data to enable a successful eco-management plan. This could involve consulting with experts in fields such as ecology, engineering, and social science to ensure that the plan is based on the best available science and is tailored to the unique characteristics of the local environment. By following these steps, it is possible to develop a comprehensive eco-management plan that promotes sustainability and protects the environment for future generations. [268]

---

268 https://www.sciencedirect.com/science/article/abs/pii/S0169204697000959

Crop rotation

Crop rotation is a practice that has been used for centuries for controlling diseases and pests in agriculture and to improve soil health and increase crop productivity. The benefits of crop rotation go beyond just improving soil quality, but also include reducing the use of synthetic fertilizers and pesticides, decreasing erosion, and supporting biodiversity. By rotating crops, different plant families are grown in a particular field in a sequential manner over time. This can help break up pest and disease cycles, as well as allow the soil to recover from the depletion of specific nutrients that certain crops may cause.[269]

In addition, crop rotation can also lead to a more resilient and sustainable agricultural system. A diverse crop rotation system can provide habitat for beneficial insects and pollinators, support wildlife, and reduce the risk of monoculture and crop failure due to weather events or pest outbreaks.

A study conducted in Iowa found that planting oats after corn or soybeans reduced the use of pesticides by 60% and the use of synthetic nitrogen fertilizer by 97%. This practice also resulted in similar yields compared to continuous corn or soybean production.[270] Some crops, like corn, are heavy nitrogen feeders and can deplete the soil of nitrogen if grown continuously in the same area. By planting leguminous crops are able to fix nitrogen from the atmosphere into the soil, reducing the need for synthetic nitrogen fertilizers. Additionally, some crops, like deep-rooted cover crops, can help improve soil structure by breaking up compacted soil layers and increasing soil organic matter. It is important to carefully plan and implement crop rotations based on local climate, soil type, and pest pressures to maximize the benefits and minimize potential drawbacks.[271] Overall, incorporating crop rotation into

---

269 https://www.harvestogroup.com/post/the-benefits-of-crop-rotation-for-soil-health-and-crop-productivity

270 rotating-crops-report-ucs-2017.pdf

271 https://www.sciencedirect.com/topics/agricultural-and-biological-sciences/crop-rotation

an eco-management plan can have numerous benefits for both the environment and agricultural productivity. By creating a more diverse and resilient agricultural system, farmers can reduce their reliance on synthetic inputs and work towards a more sustainable and healthy food system. [272]

Rotation of crops reduces pathogens, insects, parasitic nematodes, and weeds in the soil that can cause diseases and survive in the soil building up to unmanageable levels, decreasing yield. Rotation can also reduce pests and reliance on chemical pesticides. [273]

Researchers have shown crop rotation can effectively improve the climate resilience of crops in drought condition and biodiversification disease-resistant varieties of rice, soybean, corn, wheat, corn, chickpea, quinoa, buckwheat, beans, rice, watermelon, eggplants, and peppers. First, developing crop rotation technology based on local conditions; second, paying attention to the ecological benefits of crop rotation subsidies, followed by implementing appropriate and flexible subsidy policies; and, finally, carrying out rational evaluations and policy adjustment of crop rotation practices.[274]

Crops that actually suppress a disease may do so by encouraging diversity of soil organisms that outcompete or consume plant pathogens. Vegetables including onion, broccoli, beets, cabbage, kale beans, tomato, squash carrots collard and turnip greens and some grains can improve the soil through crop rotation.

Crop production is constantly under threat from weeds and insects, which can cause significant damage to crop and ultimately lead to reduced yields. Weeds can grow rapidly and consume the resources that crops need to grow and develop properly. Insects, on the other hand, can eat the crops or transmit diseases, leading to significant losses in yield. Chemical pesticides were

---

272 https://foodprint.org/issues/pesticides/
273 https://www.farmprogress.com/management/crop-rotation-as-a-method-of-disease-control
274 https://www.mdpi.com/2073-4395/12/2/436

initially developed to address this issue and have been used for thousands of years. However, the modern use of pesticides dates back to the 1940s with the introduction of DDT and weed killers like 2,4-D. DDT was widely used for decades until its negative impacts on the environment and human health were discovered, leading to its ban in many countries. This led to the development of newer pesticides like Glyphosate and Atrazine, which are now widely used. However, the use of pesticides has led to the development of pesticide-resistant weeds and insects, leading to the need for more and stronger pesticides. The result is that today there are over a thousand different pesticides being applied to crop production, many of which have negative impacts on the environment and human health. The use of eco-friendly alternatives such as crop rotation, integrated pest management, and biological control can reduce the dependence on chemical pesticides and help protect the environment and human health. [275]

However, achieving a more environmentally friendly approach to farming can be difficult in the current agricultural landscape. The use of mega-farms and industrial-scale farming has led to the prioritization of profits over environmental concerns. Farmers are often pressured to maximize yields and minimize costs, which can result in the excessive use of pesticides and other harmful practices.

To address this issue, policymakers and consumers can play a role in promoting sustainable farming practices. This can involve providing incentives for farmers to adopt environmentally friendly practices, supporting research and development of alternative pest management strategies, and educating the public about the benefits of sustainable agriculture. Consumers can also choose to support local farmers who prioritize sustainability and eco-friendly practices and demand more transparency in the food production process. By working together, we can create a more sustainable and harmonious relationship between agriculture and the environment.

---

275 https://gro-intelligence.com/insights/a-look-at-fertilizer-and-pesticide-use-in-the-us

Organic fields

In addition to the land requirements, there are other regulations that farms must comply with to be eligible for organic certification. For example, the organic regulations also specify that organic farmers must use only approved substances and practices for pest management, fertilizer use, and soil management (§205.203). They must also use organic seeds and planting stock unless organic seeds or planting stock are not commercially available in the quantity and quality necessary for the farmer's needs (§205.204). [276]

Furthermore, organic farmers must keep detailed records of their production and handling practices for at least five years (§205.103). [277] They must also maintain buffer zones to prevent contamination of organic crops by prohibited substances from adjacent conventional farms or other potential sources of contamination (§205.206). [278]

The USDA organic regulations also include specific rules for livestock production (§205.238). Organic livestock must have access to the outdoors and must be provided with organic feed and forage. Antibiotics and growth hormones are prohibited, and there are restrictions on the use of synthetic substances in livestock feed and veterinary practices.

Overall, the USDA organic regulations are designed to promote sustainable agriculture and protect the environment, while also ensuring the integrity of organic products for consumers.[279]

Pesticides are a common practice in conventional agriculture, but their negative impact on human health and the environment cannot be ignored. The USDA certification process for organic

---

276 https://www.ecfr.gov/current/title-7/subtitle-B/chapter-I/subchapter-M/part-205/subpart-C/section-205.203
277 https://www.ecfr.gov/current/title-7/subtitle-B/chapter-I/subchapter-M/part-205/subpart-B/section-205.103
278 https://www.ecfr.gov/current/title-7/subtitle-B/chapter-I/subchapter-M/part-205/subpart-C/section-205.206
279 https://www.ams.usda.gov/sites/default/files/media/Crop%20-%20Guidelines.pdf

farming ensures that farmers comply with strict regulations that prohibit the use of synthetic pesticides, fertilizers, and genetically modified organisms. To obtain a USDA organic certification, a field must be free of prohibited substances for three years before the crops can be sold as organic. This is to ensure that any residue from the previous use of pesticides is fully eliminated from the soil, and the soil has been restored to its natural state. The transition period is crucial as it allows the soil to rebuild its natural microbial communities, which can help prevent disease and pests naturally. The certification process is not only essential for human health but also for the environment. Organic farming practices promote soil health, biodiversity, and water conservation, resulting in a healthier and more sustainable food system.

It takes 3 years for a field being fallow from pesticides to obtain a USDA certificate. Here are the rules are for organic fields:

"Example to determine if a field qualifies to be organic:
A farmer's last fertilizer (or other prohibited substance)
application is on May 31, 2011.
On May 31, 2012, the first transition year ended.
On May 31, 2013, the second transition ended.
On May 31, 2014, the third transition year ended.
This means after May 31, 2014, crops harvested can be
certified organic " [280]

Crops can be grown in these fields; however, no pesticides can be used on these fields else they will not be eligible USDA organic. USDA does allow natural or non-synthetic pesticides. These are naturally occurring microscopic fungus used to infect and fight targeted insects. The program does allow use of synthetic pesticides under highly controlled applications. In organic handling production, nonagricultural synthetic, nonagricultural non-synthetic (natural), and nonorganic agricultural substances are only allowed if included on the National List (§§ 205.605 -

---

280 https://www.ams.usda.gov/sites/default/files/media/Crop%20-%20Guidelines.pdf

205.606)". [281] Organic farmers ranchers can use the substances below but the must have a disease and pest plan how they will prevent and manage pests without the use of National List inputs.[282]

In addition to reducing the incidence of disease, crop rotation can also benefit the soil by promoting beneficial microbes, improving soil structure, and increasing organic matter. Crop rotation can also improve nutrient cycling, as different crops have varying nutrient requirements and can help to replenish the soil with different nutrients.

It is important to note that crop rotation must be done correctly to be effective. Crop selection, timing, and frequency are important factors to consider when planning a crop rotation system. Farmers must also consider the impact of crop rotation on pest and disease management, soil fertility, and crop yield. Overall, crop rotation is an important tool in the sustainable management of agricultural systems. [283]

Bioagricert is an independent organization that was founded in 1984 with the aim of providing certification and control services to organic operators. Since its inception, the organization has been committed to promoting sustainable agriculture practices and ensuring that organic products meet strict quality standards. To achieve this, Bioagricert has developed an annual "Control Plan" that is tailored to the specific risks associated with each organic operator. The Control Plan includes a variety of measures, such as regular inspections and sampling, to ensure that organic operators are following the reference standards.

The Control Plan developed by Bioagricert is a comprehensive program that considers the unique challenges associated with organic farming. For example, since organic farming relies

281 https://www.ecfr.gov/current/title-7/subtitle-B/chapter-I/subchapter-M/part-205/subpart-G/subject-group-ECFR0ebc5d139b750cd/section-205.605
282 https://www.global-organics.com/post.php?s=2018-02-02-are-pesticides-allowed-in-organic-farming
283 https://www.farmprogress.com/management/crop-rotation-as-a-method-of-disease-control

on natural methods of pest and disease control, the Control Plan includes measures to ensure that organic operators are using approved methods for controlling pests and diseases. The Control Plan may also include measures to ensure that organic operators are following proper crop rotation practices, which can help to prevent the buildup of pests and diseases in the soil.

In addition to regular inspections and sampling, the Control Plan may also include unannounced inspections to ensure that organic operators are maintaining compliance with the reference standards at all times. This helps to prevent organic operators from simply "cleaning up" their operations in anticipation of a scheduled inspection. By including both announced and unannounced inspections, Bioagricert is able to provide a comprehensive evaluation of each organic operator's compliance with the reference standards.

Overall, the Control Plan developed by Bioagricert is an essential tool for ensuring that organic products meet strict quality standards. Through its comprehensive approach to certification and control, Bioagricert is helping to promote sustainable agriculture practices and ensure that consumers have access to high-quality organic products.[284]

Bioagricert is not only authorized to certify organic production in Italy, but also internationally. In addition to being recognized by the USDA, Bioagricert is accredited by the European Union (EU), Japan, Canada, and Switzerland. This means that products certified by Bioagricert can be sold as organic in these countries as well. Bioagricert's international recognition highlights the organization's commitment to promoting organic farming practices and ensuring the integrity of organic products.

The National Organic Program (NOP) was established in 2001 as a federal regulatory framework for organic agriculture in the United States. The NOP is responsible for developing and implementing standards for the production, handling, and labeling

---

284 https://www.bioagricert.org/en/

of organic products. These standards include criteria for soil and water quality, pest and weed control, and the use of additives and other inputs. The NOP also establishes guidelines for organic certification, which are carried out by third-party certifiers who are accredited by the USDA. [285]

The NOP regulations require organic farmers to use only approved substances and practices in their farming operations. They must maintain detailed records of their farming practices and submit to annual inspections by a USDA-accredited certifying agent. Organic food products that meet the NOP standards can be labeled with the USDA Organic seal, which indicates that the product was produced and handled according to the USDA's organic regulations.

The NOP has been instrumental in promoting the growth of the organic agriculture industry in the United States. The program has helped to establish uniform standards for organic production and certification, which have increased consumer confidence in organic products. The NOP has also encouraged the adoption of sustainable farming practices that reduce the use of synthetic inputs and promote soil health and biodiversity. [286]

Operators wishing to export to the USA can choose two different methods: One method is to apply for NOP certification and the second is apply for export under the European Union National Organic Program (EU-NOP) equivalence regime.

Export organic regulations out of the United States under USDA is National Organic Program (NOP) which is part of the Agricultural Marketing Service (AMS). To export to Canada organic growers must certify their produce are without sodium nitrate or hydroponic/aeroponic methods. The difference between ruminant

---

285 https://www.ers.usda.gov/webdocs/publications/42476/17400_aib780e_1_.pdf?v=0
286 https://www.bioagricert.org/en/certification/organic-production/nop-usa.html

animals are herbivores whereas non-ruminant animals are omnivores or carnivores. When shipping the export, it is "Certified in compliance with the terms of the US-Canada Organic Equivalency Arrangement." [287]

Produce can be called organic if it's certified to have grown on soil that had no prohibited substances applied for three years prior to harvest. Prohibited substances include most synthetic fertilizers and pesticides.

As for organic meat, regulations require that animals are raised in living conditions accommodating their natural behaviors (like the ability to graze on pasture), fed 100% organic feed and forage, and not administered antibiotics or hormones.

Organic certification requires that farmers and handlers document their processes and get inspected every year. Organic on-site inspections account for every component of the operation, including, but not limited to, seed sources, soil conditions, crop health, weed and pest management, water systems, inputs, contamination and commingling risks and prevention, and record-keeping. Tracing organic products from start to finish is part of the USDA organic promise.

When packaged products indicate they are "made with organic [specific ingredient or food group]," this means they contain at least 70% organically produced ingredients. The remaining non-organic ingredients are produced without using prohibited practices (genetic engineering, for example) but can include substances that would not otherwise be allowed in 100% organic products. "Made with organic" products will not bear the USDA organic seal, but, as with all other organic products, must still identify the USDA-accredited certifier. You can look for the identity of the certifier on a packaged product for verification that the organic product meets USDA's organic standards. [288]

---

287 https://www.ams.usda.gov/sites/default/files/media/Exporting%20Organic%20Products%20Factsheet.pdf
288 https://www.usda.gov/media/blog/2012/03/22/organic-101-what-usda-organic-label-means

Other sources for protein

Another food for thought is development of animal protein requires eight times as much fossil-fuel energy than growing plant protein. Meanwhile animal protein is only 1.4 times more nutritious for humans than comparable plant protein. [289] Using mushrooms are a rich source of protein low in calories, have fiber, B vitamins, selenium, potassium, and antioxidants. Mushrooms are not a one-to-one substitute for meat. As an example, we would need to eat a pound and a half of mushrooms to equal three ounces of chicken's protein. [290]

In addition to being a good source of protein, mushrooms have been found to have several health benefits. Research has shown that they may help lower cholesterol levels, improve immune function, and have anti-inflammatory effects. They are also low in fat and calories, making them an excellent choice for those looking to maintain or lose weight.

In recent years, there has been a growing interest in the potential of mushrooms as a sustainable and environmentally friendly source of protein. Compared to conventional animal protein, mushroom cultivation requires significantly less land, water, and other resources. In fact, it has been estimated that producing one pound of mushrooms requires just one gallon of water and generates only 0.7 pounds of CO_2 emissions, making them one of the most eco-friendly forms of protein available.

Furthermore, there are many different varieties of mushrooms, each with its own unique flavor and nutritional profile. Some of the most popular types include shiitake, portobello, and oyster mushrooms. These can be used in a variety of dishes, such as soups, stir-fries, and salads, and can be cooked in a variety of ways, including grilling, roasting, and sautéing.

---

289 https://news.cornell.edu/stories/1997/08/us-could-feed-800-million-people-grain-livestock-eat
290 https://mokufoods.com/blogs/nutrition/9-types-of-mushrooms-that-make-great-meat-replacements

In addition to being a delicious and nutritious food, mushrooms have also been used in traditional medicine for centuries. Some of the compounds found in mushrooms, such as beta-glucans, polysaccharides, and triterpenoids, have been found to have anti-cancer, anti-viral, and anti-inflammatory effects. While more research is needed to fully understand the potential health benefits of mushrooms, they are certainly a food worth considering as part of a healthy and sustainable diet.

As a comparison water consumption to raise livestock grain-fed beef production takes 1300 gallons of water for every pound of beef. Raising chickens takes 500 gallons of water to make a pound of meat. In comparison, soybean production uses 260 gallons a pound; rice uses 250 gallons a make pound, wheat uses 115 gallons a make pound, and potatoes uses 61 gallons a make pound. [291] All of these quantities of water usage are reasonable, and we can change where and how we grow these meat products.

Water is a precious resource, and the amount of water used in livestock production is a significant concern. The water usage in animal agriculture not only affects the local water supply but also contributes to the global water scarcity issue.

However, there are ways to reduce water usage in animal agriculture. For instance, using more efficient irrigation systems, changing feed practices to reduce water usage, and selecting more drought-resistant crops can all contribute to reducing the water usage in animal agriculture. In addition, choosing plant-based protein sources or incorporating more plant-based meals into our diets can significantly reduce the water usage associated with food production.

Moreover, water usage is not the only environmental impact of animal agriculture. The large amounts of manure produced by livestock can pollute water sources, and the methane emissions from livestock contribute to greenhouse gas emissions and cli

---

291 https://news.cornell.edu/stories/1997/08/us-could-feed-800-million-people-grain-livestock-eat

mate change.Overall, it is important to consider the environmental impact of our food choices and work towards more sustainable and efficient agricultural practices to protect our natural resources for future generations.

Well, if you are a meat eater as a majority of the population in the US is then vegetarians this comparison is food for thought, enjoy the pun. [292] However, with global warming and water shortages occurring we need to evaluate all farming methods available for protein in our diets. [293] Looking at a glass that is half full is looking at how many opportunities are available.

One opportunity that arises from the need to evaluate our farming methods is the development and promotion of alternative protein sources. [294] Beyond meat, for example, is a plant-based meat substitute that has gained popularity in recent years. This type of product not only reduces the need for animal agriculture but also has a lower environmental impact as it requires less water and produces fewer greenhouse gas emissions. Insects are another source of protein that is gaining traction in some parts of the world, such as Asia and Africa. [295] Insects require significantly less water and resources to produce than traditional livestock and can provide a sustainable source of protein for human consumption. Another alternative protein source is lab-grown meat, also known as cultured meat, which is produced by cultivating animal cells in a lab. [296] While still in the early stages of development, this technology has the potential to drastically reduce the environmental impact of meat production and address some of the ethical concerns surrounding animal agriculture. By exploring and promoting these alternative protein sources, we can work towards a more sustainable and equitable food system.

---

292 https://www.ipsos.com/en-us/news-polls/nearly-nine-ten-americans-consume-meat-part-their-diet
293 https://climatechange.chicago.gov/climate-impacts/climate-impacts-agriculture-and-food-supply
294 https://gfi.org/resource/alternative-proteins-for-a-sustainable-secure-and-prosperous-future/
295 https://www.theguardian.com/environment/2022/aug/24/insects-meat-flavor-mealworms-research
296 https://www.cnn.com/2023/06/23/business/lab-grown-meat-explainer/index.html

A few common pesticides

There are multiple reasons to be concerned about the abundance of chemicals being applied on crops. There are still unknown effects of pesticides on the ecosystem of birds, animals, insects, and amphibians. A staggering 209 million pounds of pesticides were used in California in 2016, according to the data released by the Department of Pesticide Regulation. [297] This is only California.

About 235 million pounds were applied on crops in 2018 according to the report compiled by World Animal Protection and the Center for Biological Diversity. [298] We have the endangered species act that is supposed to protect other living species.

As stated, there are over 1,000 different types of pesticides on the market today. Consumers are advised to look at the label before eating what is inside the package. Consumers have been informed to know what these harmful pesticides can potentially do by eating them over time.

Glyphosate is a broad-spectrum herbicide that has been widely used in agriculture, forestry, and urban environments for over four decades. Its effectiveness in controlling weeds has made it the most popular pesticide in the world, and it has become a cornerstone of modern agriculture. However, its use has also been highly controversial, and there have been concerns about its potential impact on human health and the environment.

Glyphosate residues have been found in many different types of feed, including grains, forages, and oilseeds. While glyphosate has been considered to be generally safe for animal health, recent studies have shown that glyphosate-based herbicides (GBHs) have the potential to cause adverse effects in animal reproduction. For example, GBHs have been found to disrupt key regulatory enzymes in androgen synthesis, which can lead to reduced fertility and other reproductive problems.

---

297 https://beyondpesticides.org/dailynewsblog/2018/05/two-hundred-million-pounds-toxic-pesticides-used-california-year/
298 https://reports.worldanimalprotection.org/US/collateral-damage#1

In addition, studies have shown that exposure to GBHs can cause alterations in serum levels of estrogen and testosterone, which can have further impacts on reproductive health. GBHs have also been linked to damage to reproductive tissues and impairment of gametogenesis, which can lead to reduced fertility and reproductive problems in both males and females.

Given these concerns, it is important to carefully consider the potential risks posed by glyphosate residues in animal feed. The use of GBHs in agriculture may have unintended consequences for animal agriculture, and the potential risks of these residues need to be taken seriously. It is important for regulators and industry stakeholders to work together to find safer and more sustainable alternatives to glyphosate and other pesticides, in order to protect animal health and ensure the sustainability of our food systems. [299]

In the European Union they are much more aggressive then here in the Unites States as they have banned 464 known pesticides the United States has banned 21. It would be great if our EPA and other government agencies in the United States could work together with the European Environment Agency (EEA) in solving a global banned on our pesticides problem. [300]

Chlorpyrifos, also known as Lorsban® or Dursban®, is an organophosphate (OP) insecticide that has been linked to negative effects on human health, particularly in children. [301] In addition to being banned in California in 2020, chlorpyrifos has also been banned in the European Union until 2020. While it has been used globally as an insecticide in agriculture, its use has resulted in water contamination in rivers, lakes, and seawater. Additionally, when it rains, chlorpyrifos can be carried from fields to nearby water sources, further contributing to water contamination. [302] In addition to surface water, well water can also be affected if it is exposed to chlorpyrifos. This type of well water contamination can be caused

---

299 https://www.ncbi.nlm.nih.gov/pmc/articles/PMC7386766/
300 https://pan-germany.org/pestizide/global-network-congratulates-countries-phasing-out-highly-hazardous-pesticides/?print=print
301 https://www.ncbi.nlm.nih.gov/books/NBK499860/
302 http://npic.orst.edu/factsheets/chlorpgen.html#products

by chlorpyrifos-containing products used for termite control. The negative impacts of chlorpyrifos on the environment and human health highlight the need for more comprehensive regulation and oversight of pesticide use. [303]

Chlorpyrifos use is on agricultural products and non-agricultural areas throughout the US. Crops including apples, citrus, corn, oranges, strawberries, wheat, sugar beets, and other foods people eat daily.

Chlorpyrifos is a broad-spectrum insecticide that has been used in the agricultural industry for over 50 years. [304] Despite its widespread use, it is known to be harmful to both humans and the environment. In addition to the negative impacts on human health, chlorpyrifos has been linked to the decline of several wildlife species, including the bald eagle and the California condor.

Chlorpyrifos are used on golf courses, and to control fire ants and mosquitoes for public health purposes and treat wood fences and utility poles. Causes from eating chlorpyrifos pesticide can permanently damage the developing brains of children, causing reduced IQ, loss of working memory, and attention deficit disorders. [305]

The use of chlorpyrifos has been linked to acute and chronic health effects, including respiratory distress, vomiting, and convulsions. The pesticide can also cause long-term neurological damage in children and has been associated with developmental delays and behavioral problems. Due to the potential risks associated with chlorpyrifos, several countries have banned its use, including the European Union and Canada.

The use of alternative pest control methods, such as integrated pest management and biological control, could reduce the need for harmful pesticides like chlorpyrifos. [306] Additionally, supporting sustainable and organic farming practices could help minimize the use of pesticides and promote healthier ecosystems.

---

303 https://pubmed.ncbi.nlm.nih.gov/32450436/
304 https://www.cdpr.ca.gov/docs/pressrls/2019/100919.htm
305 https://earthjustice.org/features/chlorpyrifos-what-you-need-to-know
306 https://portal.ct.gov/CAES/Fact-Sheets/Entomology/Approaches-to-the-Biological-Control-of-Insect-Pests

By reducing our reliance on harmful pesticides, we can protect both human health and the environment.

You can be exposed to chlorpyrifos by eating or breathing or getting this pesticide on your skin or in your eyes. [307] You can also be exposed to chlorpyrifos if you apply products containing chlorpyrifos as part of your job or applying on at your residence. This could include a bait station and could affect your pets.

On Dec. 14, 2022, EPA issued a Notice of Intent to Cancel three (3) chlorpyrifos pesticide products because they bear labeling for use on food and the registration number for chlorpyrifos is 93182-3. [308]

When it comes to identifying harmful pesticides, the registration numbers on application for chlorpyrifos to the EPA are not enough to fully understand how this chemical is being integrated into products. In fact, many pesticides have obscure names or multiple names under different formulations and trade names, making it difficult to track and regulate their use.

Chlorpyrifos itself is a prime example of how a product developed as a nerve agent during World War II was repurposed to kill insects. Over the years, chlorpyrifos has been applied on various crops and non-agricultural areas throughout the US, such as golf courses, fire ant and mosquito control, and treating wood fences and utility poles.

However, identifying chlorpyrifos as one product out of over 1,000 pesticides manufactured today is just a fraction of what is happening to life on earth. [309] There are numerous other pesticides that are also causing harm to the environment, wildlife, and humans. Therefore, there is a need for greater transparency and regulation in the pesticide industry, including stricter testing and labeling requirements to protect both public health and the environment.

---

307 http://npic.orst.edu/factsheets/chlorpgen.html
308 https://www.epa.gov/ingredients-used-pesticide-products/chlorpyrifos
309 https://www.epa.gov/newsreleases/epa-announces-plan-protect-endangered-species-and-support-sustainable-agriculture

Reading this information, one realizes we as consumers are dependent upon our agencies to protect the food we eat from farm to table. The Food Quality Protection Act (FQPA) and the USDA Pesticide Data Program (PDP) are important steps in protecting consumers from harmful pesticide residues in their food. [310] However, the FQPA safety standard does not apply to non-food uses of pesticides, such as in landscaping or home pest control, leaving consumers vulnerable to exposure.

Furthermore, while the PDP provides valuable information on pesticide residues in commonly consumed foods, it has its limitations. For instance, the program only tests a small subset of the pesticides approved for use in the US, and the testing may not be representative of all produce in the market.

It is also important to note that there are gaps in our knowledge of the long-term health effects of chronic low-level exposure to multiple pesticides. Some scientists argue that the current safety standards do not adequately protect vulnerable populations such as children, pregnant women, and farmworkers who are disproportionately exposed to pesticides.

Overall, while the FQPA and PDP are important initiatives, there is still room for improvement in our regulatory system to ensure the safety of our food supply and protect public health. [311]

It is important to note that the limited scope of testing under the Pesticide Data Program (PDP) poses a significant risk to public health. [312] As mentioned, there are over 1,000 pesticides on the market, and many of these are mixed together to create new chemical combinations. Without comprehensive testing, it is impossible to know the full impact of these chemical mixtures on human health. Farmers are mixing pesticides together and mixing chemicals together that can create new chemicals. Once chemical pesticides are mixed together unless there is a test designed to

310 https://www.epa.gov/safepestcontrol/food-and-pesticides
311 https://www.epa.gov/system/files/documents/2022-10/aceappendixb_pesticidedataprogram.pdf
312 https://www.epa.gov/safepestcontrol/food-and-pesticides

examine this mixture there is no new test to examine a new mix
ture to evaluate health results. Testing under the Pesticide Data
Program (PDP) [313]

In addition, the current testing methodology used by the PDP
focuses on identifying residues at levels lower than those that are
considered health risks. However, recent studies have shown that
chronic exposure to even low levels of pesticides can have harmful effects on human health, particularly in vulnerable populations
such as pregnant women and children.

Furthermore, the PDP only tests a limited number of foods
each year, leaving many other foods untested for pesticide residues. This means that consumers may be unknowingly exposed
to harmful levels of pesticides through their food.

To address these issues, there is a need for more comprehensive and frequent testing of pesticide residues in the food
supply. This could include expanding the number of foods tested,
as well as testing for a wider range of pesticides and chemical
combinations. Additionally, new testing methods should be developed to account for the complex mixtures of chemicals that are
present in many pesticides. Only by taking a more proactive and
comprehensive approach to testing can we ensure the safety of
our food supply and protect public health.

When focusing on the function of pesticides (i.e., herbicide,
insecticide, or fungicides), 46% of the mixtures contained insecticides alone, 15% fungicides alone, and 4.5% herbicides alone.
[314] Mixtures with effects associated with neurotoxicity were mainly
composed of insecticides, and most studies on the effects of fungicide mixtures (90%) were associated with effects on endocrine
regulation and/or reproduction. Dose addition was observed with
each kind of mixture except herbicide combinations. In contrast,
synergic interactions or greater-than-additive effects were mainly
reported for insecticide mixtures. [315]

---

313 https://www.nature.com/articles/s41370-022-00482-1
314 https://pubmed.ncbi.nlm.nih.gov/27312199/
315 https://hal.inrae.fr/hal-02637756/document

The search for the effects of combinations of chemicals in mammals is an important area of research, as it can shed light on the potential risks posed by exposure to multiple chemicals in the environment. Many peer-reviewed studies identified by the researchers provide a valuable starting point for understanding the potential for chemical combinations to produce synergistic effects, which are greater than the sum of the effects of the individual chemicals. [316]

The six studies that provided quantitative estimates of synergy are particularly noteworthy, as they provide concrete evidence of the phenomenon. [317] The fact that the magnitude of synergy at low doses varied from 1.5 to 3.5 is also significant, as it suggests that the effect of chemical combinations can be quite substantial even at relatively low levels of exposure. [318]

However, it is important to note that the occurrence of synergy was observed only at higher doses in other positive studies. This highlights the complex nature of the relationship between chemical exposure and health outcomes and underscores the need for continued research in this area.

Overall, the findings of this research suggest that the potential for synergistic effects from chemical combinations should be taken seriously, and that further study is needed to fully understand the risks posed by exposure to multiple chemicals.

Presented are results from cocktail combination mixing of two and this is the result on mice. Consider the potential effects on humans as you read these results.

The impact of pesticides on living organisms is a growing concern, and recent studies have shown that the effects of a mixture of pesticides can be more significant than those of individual pesticides alone. For instance, when tested separately, the pesticide did not show any noticeable impact. However, when tested together, the commercial formulation of Emulsifiable Concentrate (ES) and Microencapsulated (MP) resulted in sustained decreases

316 https://www.ewg.org/foodnews/summary.php
317 https://www.ncbi.nlm.nih.gov/pmc/articles/PMC6470005/
318 https://www.sciencedirect.com/science/article/abs/pii/S0009279716302198

in motor activity levels for dopamine and metabolites and dopamine turnover increased, as well as reductions in tyrosine hydroxylase immunoreactivity. [319] Furthermore, exposure to the mixture of pesticides had a more significant toxic effect on bone marrow preB and IgM(+) B-cell populations compared to the individual herbicides. It's important to note that the sequence of exposure to OP that elicits toxicity through a common mechanism of action can significantly influence the cumulative action at the targeted site (AChE and consequent functional toxicity), but no information is available on the nature of the interaction. Finally, cumulative neurotoxicity was found to be greater when exposed to the combination of Paraquat PQ and Maneb MB than when exposed to either pesticide alone. These findings indicate that the impact of pesticide mixtures should be taken seriously, and that further research is needed to fully understand their potential effects on the environment and human health. [320]

Note this is only two different chemicals being combined as a cocktail and these are some of the results posted from testing. Glyphosate is the most used chemical unfortunately there is little know when it is mixed as a cocktail with other pesticides, herbicides, or fungicides. [321] Again, we as consumers need to ask what happens when these chemical cocktails are mixed together as this happens often on farms applying pesticide herbicides and fungicides.

"Previous studies have shown that both food allergies and environmental pollution are increasing in the United States," said Dr. Jerschow. "The results of our study suggest these two trends might be linked, and that increased use of pesticides and other chemicals is associated with a higher prevalence of food allergies." [322]

---

319 https://www.nature.com/articles/npp2017208
320 https://hal.inrae.fr/hal-02637756/document
321 https://www.sciencedirect.com/science/article/pii/S2667010021001281
322 https://www.sciencedaily.com/releases/2012/12/121203081621.htm

Testing pesticides

Researching pesticide equipment Perkin Elmer popped up in an internet search and they have some ingenious systems for testing food. Perkin Elmer for example can test an apple using a Gas chromatography/mass spectrometry (GC/MS) methods are the traditional analytical method for pesticide residues and are widely used because of the sensitivity and selectivity offered. [323] (and you are scratching your head as to this type of equipment). Pesticide Residues Analysis by Clarus GC Application

As the use of pesticides in agriculture continues to increase, the need for reliable testing methods to monitor pesticide residues in food is becoming more critical. Recently, while researching pesticide equipment, I came across Perkin Elmer, a company that specializes in developing innovative systems for testing food. Their systems are designed to detect even trace amounts of pesticides, ensuring that food is safe for consumption.

Perkin Elmer's Gas chromatography/mass spectrometry (GC/MS) methods are one of the most traditional analytical methods for pesticide residues and are widely used due to the sensitivity and selectivity offered. GC/MS works by separating individual compounds within a sample using gas chromatography, followed by the identification and quantification of the compounds using mass spectrometry. This method is particularly useful for detecting trace amounts of pesticide residues in food samples, making it an essential tool in ensuring food safety. [324]

If you're unfamiliar with GC/MS, it's a complex piece of equipment that requires expertise to operate. However, Perkin Elmer's systems are user-friendly and can test an apple in just a few minutes. Their innovative approach to food testing is essential in today's world, where the use of pesticides is prevalent and monitoring their residues is critical to ensuring food safety. [325]

323 https://resources.perkinelmer.com/docs/app
324 https://www.perkinelmer.com/category/pesticides-in-food?utm_source=
325 https://www.perkinelmer.com/category/i-am-safety-quality file:///C:/Book%203/
app_pesticide_residues_analysis_by_clarus_690_gc_and_sq_8_application_
note_014522_01.pdf

Testing apples Perkin Elmer offers two systems to test pesticide residues in an apple. In their PDF Perkin Elmer can test for 19 different chemical compounds. [326] First, why in the heck are there this many pesticide chemicals being tested on an apple we would eat. Second, this equipment test is for tolerance level that will not affect children's health. How do we, the consumer, know what level is acceptable to eat apples that have pesticide in the appeal. Third this is a one-off test of the apple we are not testing to see if there is a cumulative amount being absorbed in the human body. This test is to check and see if the apples can go to market.

As the use of pesticides continues to grow, there is an increasing need for companies that can test food for pesticide residues. Multiple companies offer equipment and lab services to comply with regulatory pesticide analysis in foods. These companies play a critical role in ensuring that food is safe for consumption and that humans are protected from pesticide poisoning.

One of the companies that offer pesticide testing equipment is 908devices, which provides handheld and portable mass spectrometry devices for real-time, on-site analysis. ThermoFisher Scientific also offers analytical instruments, including gas chromatography-mass spectrometry (GC-MS) and liquid chromatography-tandem mass spectrometry (LC-MS/MS), for pesticide residue analysis. [327]

Another company that offers pesticide testing services is Agilent, which provides a range of analytical instruments and software for pesticide analysis. PrimusLabs is a USDA-accredited laboratory that offers pesticide residue testing services to growers, shippers, and retailers. Avazyme and AGQ Labs are other laboratories that offer pesticide residue testing services.

---

326 https://www.perkinelmer.com/category/pesticides-in-food?utm_source=
327 https://www.thermofisher.com/us/en/home/industrial/food-beverage/food-analytical-testing/pesticide-residues-analysis.html

Eurofins is a global scientific testing company that provides a wide range of testing services, including pesticide residue analysis. The company has a network of laboratories around the world that can test food samples for pesticide residues, ensuring that they comply with regulatory requirements.

These are just a few of the many companies that provide equipment and lab services to test for pesticides in food. Their work is essential in protecting human health and ensuring that food is safe for consumption.

The release of The Poison Papers in 2017 was a significant event in the fight for transparency and accountability in the chemical industry. The Poison Papers are a collection of over twenty thousand records obtained from federal agencies and chemical manufacturers, revealing that both regulators and chemical companies were hiding information from the public. [328]

In addition to The Poison Papers, there is also Toxicdocs, [329] which contains millions of pages of previously classified documents on industrial poisons. Together, these records paint a disturbing picture of the callousness of some companies, including Monsanto, towards public health and safety.

One example of this callousness is a meeting held by Monsanto in 1969, where they discussed how to deal with the crisis of their PCBs (polychlorinated biphenyls) destroying the world environment. At the meeting, Monsanto executives outlined three plans to deal with the crisis. The first plan was to "go out of business," which was not a realistic option for Monsanto. The second plan was to "sell the hell out of them [PCBs]," which is precisely what Monsanto did. They sold as many of the "forever chemicals" as they could before they were banned. The third plan was to "try to stay in business in controlled applications." This plan involved hiding the facts from the public and governments so that they could continue to sell PCBs or forever chemicals until 1977 when they were banned.

---

328 https://www.poisonpapers.org/the-poison-papers/
329 https://www.toxicdocs.org/

The callousness of Monsanto's actions is apparent, and their actions had severe consequences for public health and the environment. The release of The Poison Papers and Toxicdocs is a critical step in holding companies accountable and ensuring transparency in the chemical industry. It is essential to continue to push for transparency and accountability to prevent similar situations from happening in the future.[330]

The frighting fact of PCBs "forever chemicals," continually stay in the environment forever and data show PCBs can change the development of children's brains constantly.[331]

In addition to the above two papers there is the Monsanto Papers or Monsanto's secret documents. These documents were obtained via the process called Discovery. In a civil procedure Discovery allows all parties to become familiar with evidence meaning both sides must disclose all the evidence. Monsanto people had ghost writers and colluded with the Environmental Protection Agency (EPA) and hid information on how humans absorb Glyphosate into their bodies. [332]

District Judge Vince Chhabria ruled that certain documents obtained by plaintiffs in the Monsanto Roundup multidistrict (MDL) litigation could be unsealed.

"The terms glyphosate and Roundup cannot be used interchangeably, nor can you use "Roundup" for all glyphosate-based herbicides anymore. For example, you cannot say that Roundup is not a carcinogen...we have not done the necessary testing on the formulation to make that statement." [333]

This is highly disappointing for Monsanto decision to hide information from the public as well as the EPA not to disclose the dangers of glyphosate and Roundup to the public because profits of the company outweighed protection of humans.

---

330 https://www.toxicdocs.org/
331 https://biotech.ucdavis.edu/news/original-forever-chemicals
332 https://www.wisnerbaum.com/toxic-tort-law/monsanto-roundup-lawsuit/monsanto-papers/
333 https://wakeup-world.com/2018/09/19/glyphosate-is-clearly-carcinogenic-and-monsanto-hid-the-evidence/

EPA classifies over 1,000 different pesticides either chemical or biological agents use to kill living organisms. Pesticides are used to increase crop production however, hiding facts from the public as to toxicity on humans is at the forefront of regulations. The Trump administration decimated the EPA by cutting their budget by twenty six percent in 2020. Crop production in the US uses pesticides to protect crops from insects, weeds, and infections. [334]

Despite these challenges, crop production in the US continues to rely heavily on pesticides to protect crops from insects, weeds, and infections. [335] However, there is growing concern about the potential health effects of these chemicals, especially for farmworkers and people living near agricultural areas.

Regulators are working to ensure that pesticides are used safely and that the public is informed about the potential risks. This includes monitoring the use of pesticides and setting limits on their use to minimize the potential for harm. It is essential to strike a balance between the benefits of pesticide use for crop production and the potential risks to human health and the environment. This requires ongoing research, transparency, and collaboration between regulators, industry, and the public to ensure that pesticides are used safely and responsibly.

In 2014, George Washington University conducted a study examining urine samples from 14,395 people from all walks of life as part of the annual National Health and Nutrition Examination Survey. [336] The purpose of the study was to detect and examine pesticide biomarkers and compare exposure levels to the herbicide 2,4-D, which is an ingredient of Agent Orange, from 2001 to 2014.

Unfortunately, the results of the study revealed that exposure to 2,4-D had increased significantly over the thirteen-year period. In 2001-02, only 17% of the study participants had detectable levels of 2,4-D in their urine samples. By 2012, that number had

---

334 https://www.independentsciencenews.org/health/a-precautionary-tale-how-one-small-town-banned-pesticides/
335 https://thehill.com/policy/energy-environment/482352-trump-budget-slashes-funding-for-epa-environmental-programs/
336 https://pubmed.ncbi.nlm.nih.gov/25137033/

risen to almost 40%. This increase in exposure is concerning because 2,4-D has been linked to a range of health problems, including leukemia in children, birth defects, and reproductive problems.

The rise in 2,4-D exposure levels is particularly alarming because it is a widely used herbicide in agriculture and landscaping. It is used to control weeds in crops such as corn, soybeans, and wheat, as well as in residential settings to control weeds in lawns and gardens. Despite its widespread use, there is growing concern about the potential health risks associated with exposure to 2,4-D.

Regulators are working to address these concerns by setting limits on the use of 2,4-D and other pesticides and by requiring safety testing before these chemicals are approved for use. However, the rise in 2,4-D exposure levels highlights the need for continued research and monitoring to ensure that these chemicals are used safely and do not pose a threat to human health.

One third of this group study (32.52%) were considered to have high exposure, while the remaining two thirds in this study were considered to have low or no exposure. [337]

Melissa Perry's statement at George Washington University highlights the need for further research on the potential health effects of exposure to 2,4-D, particularly during early stages of life when the body is still developing. This is important because pesticides, including 2,4-D, have been linked to a range of health problems, such as cancer, birth defects, and reproductive issues. However, there is limited knowledge on how exposure to multiple pesticides simultaneously could affect human health.

It is also concerning that the use of 2,4-D in agriculture has increased by 67% in just eight years, indicating a potential lack of regulatory oversight and caution. [338] With the significant increase in pesticide use, it is crucial to ensure that there is adequate testing and safety measures in place to protect not only human health

---

337 https://ehjournal.biomedcentral.com/articles/10.1186/s12940-021-00815-x
338 https://www.theguardian.com/environment/2022/feb/09/toxic-herbicide-exposure-study-2-4-d

but also the environment. Ultimately, prioritizing profits over public health and the environment is not a sustainable approach, and there is a need for more comprehensive and protective regulations.

Three other studies in Switzerland they did not detect the presence of glyphosate in the surface water samples. In Mexico detectable levels of glyphosate were present in practically all water samples analyzed higher than legally allowed. Finally, in a Germany study showed the presents of glyphosate in 23 of the 39 samples analyzed. [339]

A questionary was created to estimate the pesticide residual intake from fruits and vegetables with urinary concentrations biomarker. Using a pesticide residue burden score (PRBS see appendix for more information) and frequency and surveillance of food pesticide residues exposure in the past year.

Two urine samples were analyzed for seven biomarkers of organophosphate and pyrethroid insecticides, and the herbicide 2,4-dichlorophenoxyacetic acid. The study showed more than ninety percent of the US population has detectable concentrations of pesticide biomarkers in their urine or blood. [340]

The International Agency for Research on Cancer (IARC), a specialized agency of the World Health Organization (WHO), is responsible for evaluating the carcinogenicity of various substances. In 2015, IARC evaluated five Organophosphate pesticides, including malathion and diazinon. Following their assessment, IARC classified malathion and diazinon as probably carcinogenic to humans (Group 2A). This classification indicates that there is sufficient evidence to suggest a potential association between exposure to these chemicals and cancer in humans. Additionally, tetrachlorvinphos and parathion were classified as possibly

---

339 https://www.ncbi.nlm.nih.gov/pmc/articles/PMC8622992/
340 https://www.ncbi.nlm.nih.gov/pmc/articles/PMC5734986/

carcinogenic to humans (Group 2B), indicating that there is limited evidence of a potential link between exposure to these chemicals and cancer. The classification of these pesticides as Group 2A and Group 2B underscores the need for further research to better understand the potential risks associated with their use, and to develop strategies to reduce exposure to these chemicals.[341]

This story started out about one person who is allergic to pesticides. What pesticides I am allergic to is not clear. However, here is one study showing 90% of Americans have detectable levels of pesticides in their urine or blood, which should concern each and every one of us.

90%

of Americans have detectable levels of pesticides in their urine or blood.

341 IARC Monographs Volume 112: evaluation of five organophosphate insecticides and herbicides

The future outlook

The news is not all doom and gloom for pesticides applied on crops. The European Union 27 nations are overhauling their farm to fork policies. Europe's crop protection industry (ECPA) has adopted a set of commitments to support Europe's new Green Deal, including to invest of over €14 billion in new technologies and more sustainable products by 2030. By overhauling these 27 nations farm and fork is to reduce pesticides use 50% by 2030. Plus create 25% of the farmland to be organic. [342]

We need to have better global policies on pesticides and encourage organic use of pesticides. The natural environment on our planet has gone through various changes and has evolved over millions of years, leading to the development of various species that exist today. However, the use and overuse of pesticides have had significant impacts on the environment and the species that inhabit it. It is important for us to understand and acknowledge the effects of pesticides on the environment and on us. It is not only the responsibility of scientists and researchers, but also of every individual to be aware of the impact of pesticides and to take action to protect the environment. Sharing information about pesticides and their effects is crucial for raising awareness and taking steps towards protecting our environment for the future.

The key mission of the Protected Areas is to guarantee healthy habitats: EUROPARC is committed to promoting the key contribution that protected areas make to human health, with the campaign of Healthy Parks, Healthy People.

We need to recognize here in the United States that there is a connection between Farm-to-Fork approach similar to the farm-to-fork approach being implement today. The United States is lagging behind because big companies lobby on their behalf not the behalf of consumers. There is a requirement to have healthy people, a healthy society, and a healthy planet.

---

342 https://www.europarc.org/european-policy/farm-to-fork-protectes-areas/

The war in Ukraine 2022 -2023 has endangered grain dependency in North Africa and the Middle East regions. Effects on food security for these nations in this region can or will cause social unrest and conflicts without a steady food. In addition, the increase in food prices affects the most venerable people in society. A broader solution is needed to allow war or other natural disasters to not affect farm to fork supply chains. [343]

Some of these companies that have been producing pesticides are now recognizing that we need to have health food being produced from farm to fork.

Bayer has made a commitment to reduce Plant Protection Products (PPP) by 30 percent by 2030 in the European Union. They are developing a black pepper oleoresin as a biological seed treatment on corn to reduce or replace the use of existing synthetic pesticides. One treatment of a modified pepper oleoresin is just that one new treatment. [344] As there are over 1,000 pesticides used on crops today only 999 more organic pesticides need to be developed.

Corteva Agriscience Eric Dereudre wants to create a better understanding and trust with "consumers by improving efficiency and sustainability of food production". [345] Trust needs to be built from the beginning; this is one more company who wants to make a statement.

Two companies out of the top twenty make a statement they want to make a change and help make food safe to eat and not harm the environment. Sixty years ago, the author of Silent Spring Rachel Carson provided a warning to the world. Pesticides manufacturing have become a gigantic industry dominated by twenty mega companies controlling 95% of the market worldwide. Only two companies say they want to change the way they do business and help save the planet.

---

343 https://www.slowfood.com/the-farm-to-fork-strategy-is-the-future-of-the-eu-food-system/
344 https://www.politico.eu/sponsored-content/smart-and-sustainable-food-systems/
345 https://news.agropages.com/News/NewsDetail---37633.htm

Starting with the largest company Syngenta Corp which has over thirty subsidiary companies making products for farm-to-fork. [346] Bayer corporation with over 400 companies could make a statement on how their company will develop safe pesticides to be applied to crops and how they will protect the environment moving to 2030.

It is not necessary to ridicule these mega pesticide companies because each company needs to make an action plan and be transparent with the consumer on how they will not destroy the environment and how they will protect our food system.

As our global population increases toward 8.5 billion by 2030 unfortunately pesticide use is also expected to increase globally.

Our planet at risk with 168 countries that have overuse pesticides with 64 percent of land used globally food crops are at risk of pesticide pollution. One third of these land areas are considered high risk. [347]

Reducing pesticide pollution is crucial for protecting biodiversity and maintaining soil health, which in turn contributes to food security. Pesticides not only affect the targeted pests but can also harm non-target organisms, including beneficial insects, birds, and soil microorganisms that play a crucial role in maintaining soil fertility and health. Exposure to pesticides can also lead to the development of pesticide-resistant pests, which can cause greater crop damage and lead to increased pesticide use.

Furthermore, a recent report called Collateral Damage highlights the impact of meat and dairy consumption on the environment and animal health. The report suggests that consumers may be contributing to the use of harmful pesticides indirectly by consuming products from animals that were raised on pesticide-contaminated feed. [348]

---

346 https://en.wikipedia.org/wiki/Syngenta
347 https://www.downtoearth.org.in/news/agriculture/64-of-world-s-arable-land-at-risk-of-pesticide-pollution-study-76206
348 https://dkt6rvnu67rqj.cloudfront.net/sites/default/files/media/WAP_Collateral_Damage_Report_02_04_22_R3.pdf

In addition, a study on infertility treatment found that consuming high-pesticide residue fruits and vegetables was associated with a lower probability of live birth, highlighting the potential health impacts of pesticide exposure on human health.

To address these concerns, California has been collecting annual pesticide use data and evaluating pesticides for their adverse impacts on human health and the environment before registering them for sale or use. DPR continuously assesses registered pesticides based on new information to mitigate potential health and environmental impacts.

The Pesticide Data Program (PDP), initiated in 1991, focuses on measuring pesticide residues in foods that are important parts of children's diets, including apples, apple juice, bananas, carrots, grapes, green beans, orange juice, peaches, pears, potatoes, and tomatoes. Samples are collected from food distribution centers in 10 states across the country. Different foods are sampled each year and then analyzed in various state and federal laboratories for the presence of residues of about 300 pesticides and similar chemicals. [349]

We need to develop an eco-management plan that requires careful consideration of various factors. The first step is to identify the social values and priorities that should be considered. Second, protecting natural resources, preserving cultural heritage, and promoting economic development. Once these priorities have been established, it is important to create clear boundaries and spaces for the project to ensure that it stays within its intended scope.

The third step involves identifying the social benefits that will be achieved by the project. These could include job creation, improving the quality of life for local communities, and promoting sustainable development.

The fourth step involves identifying and addressing any potential strains or tensions that may arise during the project. This

---

349 https://www.epa.gov/system/files/documents/2022-10/aceappendixb_pesticidedataprogram.pdf

could include conflicts over resource use, differing opinions on how to achieve goals, or concerns about the impact of the project on the environment.

The fifth step is to consider the biological results that can be achieved through the project. This may involve identifying and protecting endangered species, improving habitat quality, or reducing pollution levels.

The sixth step is to assess the benefits and costs of the project and determine if it is sustainable over the long term. This could include evaluating the economic, social, and environmental impacts of the project and identifying any potential trade-offs or unintended consequences.

The seventh step involves gathering all the relevant scientific data to enable a successful eco-management plan. This could involve consulting with experts in fields such as ecology, engineering, and social science to ensure that the plan is based on the best available science and is tailored to the unique characteristics of the local environment. By following these steps, it is possible to develop a comprehensive eco-management plan that promotes sustainability and protects the environment for future generations. [350]

We need to have better global policies on pesticides and encourage organic use of pesticides. The natural environment on our planet has gone through various changes and has evolved over millions of years, leading to the development of various species that exist today. However, the use and overuse of pesticides have had significant impacts on the environment and the species that inhabit it. It is important for us to understand and acknowledge the effects of pesticides on the environment and on us. It is not only the responsibility of scientists and researchers, but also of every individual to be aware of the impact of pesticides and to take action to protect the environment. Sharing information about pesticides and their effects is crucial for raising awareness and taking steps towards protecting our environment for the future.

---

350 https://www.sciencedirect.com/science/article/abs/pii/S0169204697000959

Mono farming

Monoculture agriculture and mono livestock farming have negative impacts on the environment and human health. [351]

The depletion of soil nutrients, water scarcity, loss of biodiversity, and the spread of disease among crops and livestock are some of the problems associated with these practices. [352] It is necessary to promote sustainable agriculture practices such as crop rotation, agroforestry, and natural pest management techniques. On the other hand, livestock farmers need to ensure that the feed they provide to their animals is nutritionally balanced and meets their specific dietary needs while considering the environmental impact of their feed production. Although the current model of industrialized animal agriculture contributes to food and nutritional security for nearly 1.3 billion people worldwide, it is not sustainable in the long term. [353] Family farmers and ranchers have an intimate knowledge of the land they work on and often adopt conservation practices that promote biodiversity and protect natural resources. They contribute to the local economy and play a crucial role in sustaining the environment. Therefore, it is necessary to find realistic solutions that work for both farmers and the environment.

Monoculture agriculture is a farming practice that involves growing only one crop on a large area of land or livestock farming has negative impacts on the environment and human health. This practice can be efficient in terms of crop production and costs, but it has a number of negative environmental and health impacts.

One of the main problems with monoculture agriculture is that it can lead to soil degradation and depleting soil nutrients, water scarcity, loss of biodiversity, and the spread of disease among

351 https://ec.europa.eu/research-and-innovation/en/horizon-magazine/rise-and-fall-monoculture-farming
352 https://greentumble.com/advantages-and-disadvantages-of-monoculture-farming
353 https://pastoralismjournal.springeropen.com/articles/10.1186/s13570-021-00223-3

crops and livestock. [354] When the same crop is grown in the same place year after year, the soil can become depleted of nutrients. This can lead to lower crop yields and the need for more fertilizer. Fertilizer use can pollute waterways and eventually contribute to climate change.

Monoculture agriculture can also lead to the loss of biodiversity and can also increase the risk of pests and diseases that target that crops can obtain. When large areas of land are dedicated to a single crop, there is less habitat available for other plants and animals. Lead farmers to use more pesticides, which can harm the environment, human health, and decline of pollinators, such as bees, which are essential for crop production.

While livestock farmers need to ensure that the feed, they provide to their animals is nutritionally balanced and meets their specific dietary needs while considering the environmental impact of their feed production. The current model of industrialized animal agriculture contributes to food and nutritional security for nearly 1.3 billion people worldwide, it is not sustainable in the long term. In 2021 the livestock emissions represented an estimated 7.1 GT of CO2-equivalent per year. Or 14.5% of human-induced greenhouse gas emissions. [355]

There are a number of things that farmers and policymakers can do to reduce the negative impacts of monoculture agriculture. These include:

Rotating crops: This helps to keep the soil healthy and reduce the buildup of pests and diseases.

Using cover crops: Cover crops help to improve soil health and prevent erosion.

Managing pests and diseases organically: There are a number of organic methods for managing pests and diseases, such as crop rotation, companion planting, and biological control.

Supporting local farmers: Buying food from local farmers helps to reduce the environmental impact of food production.

---

354 https://www.challenge.org/knowledgeitems/the-dangers-of-monocul
355 https://www.worldbank.org/en/topic/agriculture/brief/moving-towards-sustainability-the-livestock-sector-and-the-world-bank

By taking these steps, we can help to make monoculture agriculture more sustainable and protect the environment, humans, and pollinators.

Family farmers and ranchers also understand the importance of sustainable agriculture practices, such as crop rotation, cover cropping, and integrated pest management, that not only protect the environment but also maintain the health of the soil. They are often early adopters of conservation practices that help preserve natural resources, such as water and soil, while also promoting biodiversity.

Furthermore, family farmers and ranchers also contribute to the local economy, providing jobs and income for their communities. They often sell their products directly to consumers through farmers' markets, CSAs (Community Supported Agriculture), and other local markets, which helps to keep money circulating within the community.

In summary, family farmers and ranchers play a vital role in agriculture and in sustaining the environment. They are often the original custodians of the land, and their practices help to promote a healthy and sustainable food system that benefits both people and the planet.

In conclusion, while mono-agriculture may be efficient in increasing crop production and reducing costs, it comes with significant environmental and health risks. It is important for farmers and policymakers to explore more sustainable agricultural practices that reduce reliance on pesticides and promote biodiversity. The Netherlands and the EU provide good examples of how sustainable agriculture practices can be implemented, and the US should look to their examples in creating more stringent regulations on pesticide use.

Netherlands: a solution for the United States

Providing a solution in using pesticide and today is no utopia. However, the Netherlands a country the size on Maryland is second to the United States in exporting crops and livestock. They supply one third of the food to 27 European countries, but it is also a leading country in sustainable agriculture practices. [356] Dutch farmers have been using innovative techniques such as precision agriculture, vertical farming, and aquaponics to increase productivity and reduce environmental impacts. Precision agriculture, for example, uses technology such as GPS and sensors to analyze soil conditions and adjust crop management practices, accordingly, resulting in less waste of fertilizers and pesticides. Vertical farming is a technique that allows crops to be grown in vertical layers using artificial lighting and a controlled environment, which reduces the need for pesticides and water. [357] Aquaponics combines aquaculture and hydroponics, where fish and plants are grown together in a symbiotic system that produces both food and fertilizer with minimal environmental impact. [358]

Vertical farming is an innovative way of growing crops that is gaining popularity in the Netherlands and other countries. By utilizing a vertical space, these farms can produce high yields of fresh and healthy produce, while minimizing the use of land and water.

The former IBM facility that has been converted into a vertical farm is just one example of how this technology is being used in the Netherlands. Similar facilities can be found throughout the country, with many more being built every year.

---

356 https://www.weforum.org/agenda/2019/11/netherlands-dutch-farming-agriculture-sustainable/
357 https://www.edengreen.com/blog-collection/what-is-vertical-farming
358 https://gogreenaquaponics.com/blogs/news/what-is-aquaponics-blog

One of the biggest advantages of vertical farming is that it allows for year-round production of crops. This means that fresh produce can be available to consumers even during the winter months, when traditional farming is not possible.

Another advantage of vertical farming is that it can be done close to urban areas, reducing the distance that food has to travel and making it more sustainable. In addition, since vertical farming is done in a controlled environment, it is less susceptible to weather conditions and pests, which can lead to higher crop yields and less waste.

The technology used in vertical farming is also evolving rapidly, with new innovations and advancements being made all the time. For example, some vertical farms are now using artificial intelligence to optimize plant growth and monitor crop health. Others are experimenting with different types of lighting, such as LED or ultraviolet light, to see how it affects crop yields and quality.

Overall, vertical farming has the potential to revolutionize the way we grow and consume food. While it may not be able to replace traditional farming entirely, it can complement it and help to ensure that everyone has access to fresh and healthy produce.

The Netherlands' success in agriculture can be attributed to a combination of factors, including innovation, technology, and sustainability. Their emphasis on automation and vertical farming has helped to optimize land use and minimize the impact on the environment. Additionally, their focus on seed development has allowed them to produce crops that are more resistant to disease and weather conditions, leading to higher yields. The Netherlands' approach to agriculture has also prioritized reducing the use of pesticides, which has positive effects on human health and the environment.

Furthermore, the Netherlands' success in agriculture can serve as a model for other countries and farmers. By adopting similar methods, farmers can increase their productivity and decrease their environmental impact. The Netherlands' focus on sustainability is particularly relevant in the face of climate change, as it allows for food production that is both efficient and resilient.

Overall, the Netherlands' innovative and sustainable approach to agriculture has allowed them to become a leader in food production despite their small size. Their emphasis on technology and automation has made them a pioneer in vertical farming and greenhouse production, while their focus on seed development has led to higher yields and reduced reliance on pesticides. These practices can serve as an inspiration to farmers and food producers around the world, highlighting the importance of innovation, sustainability, and environmental responsibility.

The Netherlands' success in reducing pesticide use is often attributed to their innovative approach to agriculture, which involves the use of advanced technologies such as greenhouses and solar energy. [359] This has allowed them to grow crops using fewer resources, including water and fertilizers, while still maintaining high yields.

The Netherlands have limited their use of pesticides per the European Union and as of this writing EU has banned, and UK once again scored highest with 464 pesticides banned, followed by Turkey (212) and Saudi Arabia (201). [360] The United States has banned 21 pesticides, and records 81 "voluntary withdrawals" of pesticide registrations. With over 1,000 different pesticides on the market, it is nearly impossible for agencies to manage this many different products. The potential cocktail mixtures of pesticides are challenging to anyone person or agency. [361]

Furthermore, the European Union has set strict regulations on the use of pesticides to protect the environment and human health. [362] The EU's regulations are based on a risk assessment

---

359 https://www.nationalgeographic.com/magazine/article/holland-agriculture-sustainable-farming

360 https://www.panna.org/news/more-national-pesticide-bans-despite-industry-pushback/

361 https://www.panna.org/blog/more-national-pesticide-bans-despite-industry-pushback

362 https://www.europarl.europa.eu/factsheets/en/sheet/78/chemicals-and-pesticides

approach, where a pesticide is only authorized if it is shown to have no unacceptable effects on human health and the environment. In addition to banning certain pesticides, the EU also sets maximum residue levels for pesticides on food, which ensures that the food consumed by EU citizens is safe.

Additionally, the Netherlands has invested heavily in greenhouse farming, which allows farmers to grow crops year-round in a controlled environment. [363] This approach not only helps to ensure a consistent supply of fresh produce, but also reduces the need for harmful pesticides and herbicides.

Modernization to help protect the environment from pesticides and pest should be our concern as a nation. In this case should big farm groups decide to invest and follow the lead from these groups in the Netherland than I think we can correct our pesticide problem her in the United States within a decade if companies decide to invest in a solution. [364]

The United States, on the other hand, has been criticized for its lax regulations on pesticides. While some pesticides have been banned in the US, many others are still in use, and the regulatory process for approving pesticides is often criticized for being too slow and not stringent enough. In addition, there is a lack of data on the long-term effects of pesticide exposure on human health and the environment, which makes it difficult to assess the risks associated with pesticide use.

In addition to their reduction in pesticide use, the Netherlands has also been successful in reducing their carbon emissions from agriculture. This is largely due to their use of sustainable farming practices, such as crop rotation and the use of cover crops, which help to improve soil health and reduce the need for synthetic fertilizers. [365]

---

363 https://www.washingtonpost.com/business/interactive/2022/netherlands-agriculture-technology/
364 vpro.nl/foodforthought
365 https://pubmed.ncbi.nlm.nih.gov/12425108/

The Netherlands' success in sustainable agriculture has led to them being recognized as a global leader in this field. They have shared their expertise with other countries, including the United States, and have encouraged the adoption of sustainable farming practices around the world.

While there is still much work to be done in reducing pesticide use and improving sustainable agriculture practices globally, the Netherlands provides a valuable example of what can be achieved through innovative thinking and a commitment to environmental stewardship. In a 2020 study pesticide use from 2016 application decreased for most crops. [366]

Despite being a relatively small country, the Netherlands has managed to become a major player in the global food export market, trailing only behind the United States. One of the key reasons for the Netherlands' success in this area is their adoption of modern farming techniques, including the use of robotics and greenhouses.

For example, in the Netherlands, chickens are typically raised in indoor pens that allow them to move around freely while still being protected from predators and inclement weather. [367] These chickens are typically fed both mechanically and by hand to ensure that they receive the right balance of nutrients.

Similarly, cows in the Netherlands are often milked using modern milking machines, which can quickly and efficiently place them on a serviceable milking system. This allows farmers to effectively manage their herds, leading to higher milk yields and better animal health.

The Netherlands is the largest importer of soy in Europe, bringing in a staggering 9 billion tons per year. Of this, 43% is used for animal feed. [368] This is particularly important in the Netherlands, which has a significant dairy industry. The country

366 https://www.cbs.nl/en-gb/news/2022/02/less-pesticide-used-in-agriculture
367 https://www.theguardian.com/environment/2018/jun/23/living-next-door-to-17-million-chickens-we-want-a-normal-life
368 https://www.cbs.nl/en-gb/news/2020/40/soybean-imports-from-brazil-up-by-40-percent

is home to 4 million cows, with 40% of them being dairy cows. [369] These cows produce an impressive 30 billion pounds of milk per year, much of which is used to make cheese.

To manage this significant agricultural output, the Netherlands relies heavily on automation and skilled workers to manage the flow of work. This allows for a high degree of efficiency and consistency in production, ensuring that the dairy industry is able to operate at peak efficiency.

In 2021, the Netherlands produced over 2 billion pounds of cheese, with an astounding 85% of this being exported around the world. [370] This is a testament to the quality and consistency of Dutch cheese, which is renowned for its taste and texture.

Overall, the Netherlands' large-scale dairy industry and significant cheese exports are made possible through a combination of advanced automation and skilled labor. Additionally, the country's reliance on imported soy highlights the interconnectedness of global agriculture and the need for countries to work together to ensure a sustainable and efficient food system.

Overall, by leveraging these modern techniques and technologies, the Netherlands has been able to compete successfully with larger countries like the United States in the global food export market.

The Netherlands is a major global exporter of onions, producing a staggering 4 billion pounds in 2021 alone. [371] Despite this impressive output, the average person in the Netherlands only consumes about 16 pounds of onions per year, while consumers in Africa and Asia consume five to six times more.

In fact, a whopping 80% of the onions grown in the Netherlands are exported to the rest of the world, with 84% of these onions being shipped to 130 different countries.

369 https://www.france24.com/en/tv-shows/down-to-earth/20211112-gassy-grazers-the-dutch-dilemma
370 https://longreads.cbs.nl/the-netherlands-in-numbers-2022/how-much-cheese-do-we-produce/
371 https://www.tridge.com/news/the-netherlands-exported-13-million-onions-in-the-

One key reason for the Netherlands' success in onion production is their use of modern farming practices, including the use of GPS systems in tractors to monitor prices in real-time while harvesting. This allows farmers to become not only growers, but also brokers, ensuring that they are able to maximize profits and maintain competitive pricing.

Once the onions are harvested and brought to the sorting facility, the process is almost entirely automated. The sorting system uses advanced technology to sort the onions by size, storing them in bags by size and making them ready for shipment. This not only saves time and money, but also ensures that the onions are of a consistent quality, making them more attractive to buyers around the world.

Overall, the Netherlands' commitment to modern farming practices and advanced technology has allowed them to become a major player in the global onion market, exporting the vast majority of their crop to countries around the world.

Agro Care is the largest tomato grower in Europe, with an impressive 345-acre greenhouse facility located in the Netherlands. [372] This state-of-the-art facility produces an astounding 167 million tons of tomatoes each year, making it one of the most significant tomato producers in the world.

The Netherlands as a whole is a major producer of tomatoes, with an output of 2 billion pounds per year. What's more, a staggering 80% of these tomatoes are destined for export to markets around the world.

The production of these tomatoes requires a significant amount of labor, with hundreds of people employed at the Agro Care facility alone. The use of greenhouse facilities allows for a high degree of control over the growing environment, including factors like temperature, humidity, and light. This, in turn, allows for year-round production of high-quality tomatoes, even in regions with less favorable growing conditions.

---

372 https://www.agrocare.nl/en/who-we-are

The use of greenhouses also allows for more efficient water use, as water can be recirculated and reused in the growing process. This is particularly important in regions where water scarcity is a concern.

Overall, the production of tomatoes in the Netherlands is a significant contributor to the country's agricultural sector and economy, providing jobs for hundreds of people and generating significant revenue through exports. The use of modern greenhouse facilities and advanced growing techniques allows for a high degree of efficiency and consistency, ensuring that Dutch tomatoes remain a staple of kitchens around the world.

Friesland Campina is one of the world's largest dairy companies, with a significant presence in the Netherlands. [373] The company produces a range of dairy products, including evaporated milk, sweet milk, and condensed milk. What's more, much of the production process at their facility is nearly completely automated, allowing for high efficiency and consistency in production.

This automation is particularly important given that Friesland Campina is a major exporter of dairy products. In fact, 95% of the products produced at the facility are exported to markets around the world, including the Middle East, Asia, and Europe.

The production of dairy products at this scale requires a significant number of resources, including milk from local dairy farms. However, by relying on advanced automation and technology, Friesland Campina is able to streamline their production process and produce high-quality dairy products at a competitive price point.

The company's focus on export markets reflects the interconnectedness of the global food system, as well as the growing demand for dairy products in emerging markets around the world. By meeting this demand through efficient and automated production, Friesland Campina is able to remain a significant player in the global dairy industry. Overall, the production of dairy products at Friesland Campina's facility in the Netherlands is a testament

---

373 https://www.frieslandcampina.com/

to the power of advanced automation and technology in the food production industry. With the ability to produce high-quality products at scale, and export them to markets around the world, Friesland Campina is well-positioned for continued success in the years to come.

VanDrie Group part of their operations, the group separates out male calves from the rest of the herd and sends them to the slaughterhouse. [374] This focus on meat production is an important aspect of the company's overall business model and has allowed them to capture a significant portion of the market.

Despite the Netherlands' relatively small size, Bel Group a French-Dutch food company is able to export a large amount of meat to countries throughout Europe. In fact, of the estimated 1.5 million pounds of meat produced each year, a staggering 90% is exported to markets in Germany, France, and Italy. [375]

One key reason for this success is the company's use of advanced automation technologies. By using highly efficient and streamlined processes, Bel Group is able to ensure that nothing goes to waste during the production process. This includes everything from the use of advanced slaughterhouse equipment to the precise management of animal feed and waste products.

Moreover, the company's focus on sustainability and animal welfare has helped to distinguish them from competitors in the global meat market. By prioritizing the health and well-being of their animals, Bel Group is able to produce high-quality meat products that meet the needs of discerning consumers around the world.

Bel Group is one of the world's largest cheese producers and is a multinational company that specializes in the production and distribution of various dairy products, including cheese. They are known for their well-known brands such as The Laughing Cow, Babybel, and Boursin. [376]

---

374 https://www.vandriegroup.com/en/
375 https://www.washingtonpost.com/business/interactive/2022/netherlands-agriculture-technology/
376 https://www.groupe-bel.com/en/

Overall, Bel Group success in the global meat market is a testament to the power of advanced automation and technology in the food production industry. By leveraging these tools to produce high-quality meat products, the company has been able to compete on a global scale and capture a significant share of the market. Additionally, all this food being exported is inspected by others to protect the quality and safety of the products being exported and also includes animal welfare. All labels state where the product originates and the data relevant along with date and time. This includes random checks including temperature and sometimes products are checked for taste. Many items need to be frozen and sent off to large cold storage lockers prior to shipping. Part of the safety program requires food and safety personnel to randomly check food in these lockers for safety.

While the Netherlands is known for its advanced food production and exportation capabilities, it is important to note that the country also imports a significant amount of food products each year. In 2021, for example, the Netherlands imported a total of 3.7 billion dollars' worth of meat, 2.8 billion dollars' worth of vegetables, 7.1 billion dollars' worth of fruit, and 2.3 billion dollars' worth of fish and seafood. [377]

There are several reasons why the Netherlands imports such a large amount of food each year. One key factor is the country's relatively small size and limited arable land. Despite the country's advanced farming and food production capabilities, there are simply not enough resources to meet all of the population's needs. As a result, the country relies heavily on imports to supplement its domestic food supply.

Another factor contributing to the Netherlands' high levels of food imports is its position as a major global trading hub. Due to its central location in Europe and well-developed transportation infrastructure, the country serves as a major transit point for food products coming from around the world. This includes everything from fresh produce to frozen meat and seafood.

---

[377] https://www.statista.com/statistics/591124/value-meat-and-edible-meat-offal-imported-to-the-netherlands/

Despite the high levels of food imports, however, the Netherlands remains committed to sustainable and ethical food production practices. The country is a leader in the development of innovative food technologies and is constantly working to improve the efficiency and sustainability of its food production systems. As such, while the Netherlands relies on food imports to some extent, it remains a major player in the global food production and exportation industry.

The Netherlands is not only a major player in food production and exportation, but it is also home to the world's largest seed growers. The country has long been at the forefront of seed development, with a focus on creating seeds that are higher yield and more resistant to disease, drought, and other environmental stressors.

According to the Food and Agriculture Organization of the United Nations (FAO), the Netherlands exported $1.1 billion worth of seeds in 2020, which is about 7% of the global seed market. The top five exporters of seeds in 2020 were: [378]

China: $2.9 billion

United States: $2.4 billion

Netherlands: $1.1 billion

Germany: $1 billion

France: $0.9 billion

Despite the importance of seed development in the Netherlands, it is important to note that the industry is not without its challenges. One of the biggest obstacles facing seed growers is the high cost of seed production. According to a 2021 study by the University of California, Davis, the average cost of producing a pound of tomato seed is about $10. Or a seed that used to cost less can go for 100 times that today. This is due to the complex and highly specialized nature of seed production, which requires extensive research and development, as well as precision planting and cultivation techniques.

---

378 https://www.washingtonpost.com/business/interactive/2022/netherlands-agriculture-technology/

To help overcome these challenges, many seed growers in the Netherlands have turned to automation as a way to increase efficiency and reduce costs. For example, robots are often used to plant and cultivate vegetables, helping to minimize labor costs and increase precision in the growing process.

Despite the challenges facing the seed industry, however, the Netherlands remains a global leader in seed development and production. With its long history of agricultural innovation and commitment to sustainability and efficiency, the country is well-positioned to continue shaping the future of food production for years to come.

In a 2021 report published by the Dutch Ministry of Agriculture, Nature, and Food Quality. Stated that the Netherlands used about 1.3 million tons of food waste as animal feed in 2020. The report also states that the food waste was used as animal feed and is expected to increase in the coming years. In addition to using rejected food as animal feed for pigs, the Netherlands also has a strong focus on sustainability and circular economy in their food production industry. This means that waste products from food production are often repurposed and reused in various ways. For example, leftover plant material from greenhouse farming is used as animal feed or as a source of renewable energy through anaerobic digestion.

Moreover, the Netherlands has been experimenting with innovative solutions such as insect farming as an alternative source of protein for animal feed and human consumption. Insect farms are able to produce large quantities of protein-rich insects, such as mealworms and crickets, using minimal resources and space compared to traditional livestock farming.

Insect protein is a sustainable alternative to meat protein. It is produced by breeding insects in a controlled environment, and it requires less land and water than raising livestock. The Netherlands is a leading producer of insect protein, and there are a number of companies in the country that are developing insect-based food products. Protix has a factory in Delft where it produces insect protein. One insect is black soldier flies that is high in protein. These flies are processed into a powder that can be used in a variety of food products. Another company is Hexafly in Eindhoven and produces insect protein. They grow mealworms and are high in protein. The mealworms are processed into a powder and can be used in a variety of food products.

Overall, the Netherlands' strong focus on innovation, automation, and sustainability has allowed them to become a major player in the global food industry despite their small size.

The Netherlands is using sustainability and circular economy in their food production industry. The following companies all use aeroponics and LED lighting to grow their crops. Plant Lab in Amsterdam is a vertical farm and grows a variety of leafy greens and herbs, Eden Vertical Farm is a vertical farm located in Rotterdam and grows a variety of vegetables. Utrecht vertical food group is a company that operates a number of vertical farms in the Netherlands and grows a variety of leafy greens and herbs.

The Netherlands is a major producer of fish, and the food production industry is working to reduce the amount of fish waste that is produced. One way they are doing this is by recycling fish waste into fertilizer and other products. This helps to reduce pollution and create new products from waste materials.

One company, Good Fish Company in Flevoland grows fish in recycled shipping containers. The company uses a method of farming called recirculating aquaculture systems (RAS), which allows the fish to be raised in a closed system that uses minimal water.

Banned pesticides

The issue of pesticide use in agriculture is a complex one, with many stakeholders involved, including farmers, pesticide manufacturers, regulatory agencies, and consumers. While pesticides have been crucial in increasing crop yields by protecting crops from pests and diseases, there are concerns about the potential health and environmental impacts of these chemicals.

One of the challenges in regulating pesticides is the fact that different countries have different regulations and standards for pesticide use. For example, some pesticides that are banned in one country may be legal in another. This can create confusion for consumers who are trying to make informed choices about the foods they eat.

There is also growing interest in organic farming, which avoids the use of synthetic pesticides altogether. While organic farming currently represents a small percentage of global agriculture, it is growing rapidly, driven by increasing demand from consumers who are concerned about the health and environmental impacts of pesticides.

Overall, the issue of pesticide use in agriculture is likely to remain a topic of debate and discussion for many years to come. As consumers become more aware of the potential risks associated with these chemicals, there may be increasing pressure on regulators to adopt more stringent standards and to promote more sustainable agricultural practices.

Without organizations like Pesticide Action Network (PAN) we the consumer would not have advocates helping us navigate the myriad of pesticides applied to foods we eat. The following list is from 2021 and the European Chemicals Agency (ECHA) could form a relationship to work together in banning known pesticides that are harmful.

The United States is one of the largest agricultural producers in the world, but it also uses a significant number of pesticides on its crops. The US EPA maintains a chemical database that contains information on registered pesticides, and this can be accessed by the public. However, a 2019 study found that the US lags behind other agricultural nations in banning harmful pesticides, and the lack of regulation or interest in banning pesticides has been a cause of concern.

The US Federal Register is a resource that publishes government regulations and notices, and it contains information related to pesticide regulation. The US EPA has suspended, cancelled, and restricted the use of certain pesticides, and this information is also available as a resource provided by PAN (Pesticide Action Network).

The use of pesticides in the US has been a contentious issue, with concerns about the impact of these chemicals on human health and the environment. Organizations like PAN work to advocate for the banning of harmful pesticides and to raise awareness about the risks associated with pesticide use. As the world becomes more aware of the dangers of pesticides, there has been a push to phase out the use of harmful pesticides in countries like the EU, China, and Brazil, and it remains to be seen whether the US will follow suit. [379]

It is important to note that the use of pesticides and their regulation is a complex and ongoing issue. The bans or restrictions on certain pesticides in one country may not be mirrored in other countries and may also be subject to change over time. The reasons for these bans or restrictions can vary, and may include concerns about human health, environmental impacts, or other factors.

In the case of the newly listed pesticides, it is interesting to see that many of them have been banned or restricted in multiple countries, including some major agricultural producers. This may indicate growing concerns about the safety and effectiveness of these pesticides, as well as increasing efforts to regulate their use. However, it is also worth noting that not all bans or restrictions are necessarily based on scientific evidence, and there may

---

[379] https://files.panap.net/resources/Consolidated-List-of-Bans-Explanatory.pdf

be differences in opinion or interpretation of the available data. It is important for consumers, farmers, and policymakers to stay informed about the latest research and developments related to pesticides, and to consider the potential impacts on human health and the environment. [380]

According to the USDA's Agricultural Chemical Use Program, the following pounds of pesticides were applied to plants in the United States in 2021, 2020, 2019, and 2018.

2021: 1.1 billion pounds

2020: 1.09 billion pounds

2019: 1.08 billion pounds

2018: 1.07 billion pounds

The majority of these pesticides are used on agricultural crops, such as corn, soybeans, and cotton. Pesticides are used to control pests, such as insects, weeds, and diseases, that can damage crops and reduce yields. These pesticides can have negative impacts on human health and our environment. [381]

The study published by Environmental Health found that the United States allows the use of 85 pesticides that are banned or are in the process of being phased out in the European Union, China, or Brazil. These pesticides have been linked to a range of health problems, including cancer, birth defects, and neurodevelopmental disorders. The researchers behind the study compared pesticide regulations across countries and found that the United States lags behind in banning harmful pesticides.

The study also points out that the lack of regulation in the US has resulted in a higher exposure to pesticides for Americans compared to people living in other countries. This is a major concern as there is mounting evidence that links pesticide exposure to a range of health problems. The US Environmental Protection Agency (EPA) has been criticized for its lack of action in banning harmful pesticides and for not prioritizing public health.

---

380 https://files.panap.net/resources/Consolidated-List-of-Bans-Explanatory.pdf
381 https://www.nass.usda.gov/Surveys/Guide_to_NASS_Surveys/Chemical_Use/

The study is a wake-up call for the US government and the pesticide industry to take action to protect public health. It is crucial for the EPA to review and update its regulations on pesticides to ensure that Americans are not exposed to harmful chemicals in their food and environment. There is a need for more research and investment in alternative methods of pest control that do not rely on toxic chemicals. Additionally, consumers can play an important role by choosing organic and pesticide-free products and advocating for stricter pesticide regulations. [382]

Pesticides are designed to kill pests, but they can also be harmful to humans. Exposure to pesticides can cause a range of health effects, from short-term impacts such as headaches and nausea to more serious health conditions like cancer, reproductive harm, and endocrine disruption. Pesticides can also contaminate water, air, and soil, and can harm wildlife.

Safe food

Safe food production is a necessity for all of us on this planet. The Federal Aviation Administration (FAA) started because the airplane industry grew up and multiple airplane companies were competing to be first. Pilots would provide rides to people across America for a few dollars to promote aviation. Then airplane accidents started to happen too often, and people demanded there be regulations to be installed.

Regulations developed by the (FAA) are in the Federal Aviation Regulations (FARs) these regulations are followed by all private and commercial airplanes all around the world. When a pilot speaks to a control tower in another country both the pilot and the tower speak in English. Therefore, both pilot and tower can communicate in the same language as a matter of safety.

---

382 https://www.centerforfoodsafety.org/press-releases/6535/lawsuit-challenges-epas-failure-to-obey-court-order-to-protect-endangered-wildlife-from-toxic-pesticide

In 1947 the International Organization for Standardization (ISO) was established to voluntary share knowledge on a consensus basis to establish an international standard. Today ISO group has 167 national standards bodies in the world. Their purpose is to agree on standards for innovation of products.

Today there is a need for an international organization similar to FAA or ISO standards to administer regulations on allowable pesticides to be applied on our crops and what is fed to livestock and to protect the environment and all living organisms. The creation of Federal Pesticide Administration (FPA) under the EPA or similar agency to develop management rules and guidelines. Working in conjunction with the EPA as to what types of pesticides are authorized to be manufactured and applied on crops similar to the European Union Plant Protection Products. To assist farmers and livestock growers of what can be used in the environment and what can be served in the store for consumers to purchase.

The European Union has taken a more precautionary approach to regulating pesticides, which has resulted in a higher level of protection for their EU citizens. The EU's ban on certain pesticides has been driven by the need to protect human health and the environment. They have developed a list of banned substances that are considered too dangerous for use in agriculture, while the US has been more lenient in regulating pesticides.

Furthermore, the EU has also taken steps to limit exposure to pesticides by reducing their use. For example, they have set limits on the amount of pesticide residues allowed on food and have implemented measures to reduce pesticide use in agriculture. In contrast, the US has been criticized for its reliance on pesticides and for not doing enough to protect human health and the environment.

Today the United States is lagging behind the European Union in safeguarding their citizens. The European Union continues to lead in innovation that help protect their 27-member union for harmful pesticides. The European Union has banned 464 pesticide and the United States has banned 21 pesticides.

Researching many pesticides, we have learned there are many types of pesticides that should have never been released into the market. There is not enough protection for all living things on our planet. Our world society requires a plan to invest money to safeguard our food production without continually hurting the environment and humans, birds, insects, amphibians, and other mammals with pesticides.

The issue of pesticide uses and its impact on the environment and living organisms is a complex one that requires careful consideration and a multifaceted approach. The first step is to identify the harmful pesticides and stop their use. However, this is only the beginning. We also need to invest in research and development of alternative pest control methods that are safer and more sustainable. For example, integrated pest management techniques that utilize natural predators, crop rotation, and other non-toxic methods can be effective in controlling pests without the use of harmful chemicals.

Another important aspect is to educate farmers and consumers about the dangers of pesticides and the benefits of adopting sustainable farming practices. By providing incentives and support for farmers who choose to adopt such practices, we can encourage the shift towards more sustainable and environmentally friendly agriculture.

In addition, we need to improve regulatory frameworks to ensure that new pesticides are thoroughly tested before being approved for use, and that they are regularly reviewed to ensure their continued safety. This requires adequate funding for regulatory agencies and strict enforcement of regulations.

We need to foster a culture of responsibility and accountability among pesticide manufacturers, farmers, and consumers. This means taking responsibility for the impact of pesticide use on the environment and human health, and actively seeking ways to minimize harm. By working together, we can create a safer, more sustainable future for all.

The rapid growth of the pesticide industry over the last 75 years has had significant environmental and health impacts. Although there have been some efforts to regulate the use of pesticides, government agencies have not been able to keep pace with the increasing number of pesticide applications being submitted each year. This has resulted in a lack of oversight and regulation, leading to the release of many dangerous pesticides into the market.

There is a need for a global agency that can effectively monitor all pesticide production and ensure the safety of both the environment and human health. One possible solution is the creation of a Federal Pesticide Administration that could be responsible for all pesticide regulations in the United States. Such an agency would need to work closely with manufacturers to ensure that all pesticides are tested and evaluated for safety before being approved for use.

Ultimately, the creation of a Federal Pesticide Administration and the promotion of alternative pest control methods could help to reduce the use of harmful pesticides, protect the environment and human health, and promote sustainable agriculture.

Medical risks pesticides

A 2019 review article published in the journal Environmental Research examined the link between pesticide exposure and cardiovascular disease. The review found that exposure to pesticides, particularly organophosphates, was associated with an increased risk of cardiovascular disease, including hypertension, coronary heart disease, and stroke. The article also noted that the effects of pesticide exposure on cardiovascular health may be exacerbated by other factors such as age, sex, as well as any underlying medical conditions.

It is important to note that while the evidence linking pesticide exposure to high blood pressure is growing, more research is needed to fully understand the extent of this association and the mechanisms involved. However, the existing evidence highlights the potential health risks associated with pesticide use and emphasizes the importance of continued efforts to reduce exposure to harmful chemicals in our environment.

Research has shown that certain chemicals such as PCBs, phytoestrogens, fungicides, pesticides, and other xenobiotics can disrupt brain sexual differentiation. This means that exposure to these chemicals during critical periods of brain development can alter the way the brain develops in terms of reproductive physiology and behavior. This can have a long-lasting impact on an individual's reproductive health and sexual function later in life. Animal studies have shown that exposure to these chemicals can reduce reproductive success and cause abnormalities in reproductive organs. For example, exposure to the pesticide DDT has been linked to reduced fertility in male mice, while exposure to PCBs has been associated with changes in sexual behavior in rats. These findings suggest that exposure to environmental toxins may contribute to reproductive problems in humans as well. It is important to reduce exposure to these chemicals, especially during critical periods of development such as fetal and early childhood stages, in order to minimize the risk of adverse effects on reproductive health. [383]

The environmental concerns associated with pesticides include water contamination, which can occur when pesticides run off from agricultural fields into nearby streams, rivers, and groundwater sources. This contamination can harm aquatic life and can also affect the quality of drinking water. Pesticides can also contribute to air pollution, as they can be carried in the air and inhaled by people and animals.

The health risks associated with pesticide exposure can vary depending on the type of pesticide and the level and duration of

---

[383] https://www.ncbi.nlm.nih.gov/pmc/articles/PMC2726844/

exposure. Short-term exposure to high levels of pesticides can cause acute symptoms such as headaches, dizziness, nausea, and skin irritation. Long-term exposure to low levels of pesticides has been linked to chronic health conditions such as cancer, neurological disorders, and reproductive problems.

Farm workers and people living near agricultural areas are particularly vulnerable to pesticide exposure. In addition, children and pregnant women may be at higher risk due to their developing bodies and potential for exposure during critical periods of growth and development.

There is growing concern about the impact of pesticides on pollinators such as bees and butterflies, which play a critical role in food production and ecosystem health. Pesticides can also harm other non-target species, such as birds and aquatic life.

Overall, while pesticides can provide important benefits in controlling pests and increasing crop yields, it is important to carefully consider their use and potential environmental and health impacts.

Hodgkin's disease, also known as Hodgkin lymphoma, is a type of cancer that affects the lymphatic system. The causes of Hodgkin's disease are not fully understood, but research has suggested that exposure to certain environmental toxins, including pesticides, may be a contributing factor.

Several studies have investigated the link between pesticides and Hodgkin's disease. One study conducted in Canada found that individuals who reported exposure to pesticides at work had an increased risk of Hodgkin's disease. Another study conducted in the United States found that farmers who reported using certain pesticides had an increased risk of developing Hodgkin's disease.

The International Agency for Research on Cancer (IARC) has classified several pesticides as "possibly carcinogenic to humans," including glyphosate, which is the active ingredient in the popular herbicide Roundup. Glyphosate has been widely used in agriculture and landscaping, and studies have linked its use to an increased risk of various types of cancer, including non-Hodgkin's lymphoma. While the exact mechanisms by which pesticides may contribute to Hodgkin's disease are not fully understood, it is believed that they may disrupt the immune system, which plays a role in the development and progression of the disease. Additionally, some pesticides may act as endocrine disruptors, which can also affect the immune system and contribute to the development of cancer.

Overall, while more research is needed to fully understand the link between pesticides and Hodgkin's disease, evidence suggests that exposure to pesticides may be a contributing factor in the development of the disease.

Final chapter

Writing this story, I started by sharing my allergies to pesticides in our food system. However, researching this pesticide industry was at times depressing to read about birds, and bees becoming disorientated from ingesting pesticides and frogs being castrated from pesticides. Then there is the lack of conservation to land in search of making a profit over the annihilation of native plants, insect, amphibians, honeybees, birds, and mammals. Capitalism is a great tool when used with respect toward living organisms.

Obfuscation of submitting paperwork from pesticide companies to government agencies by using cryptic acronyms where only the sender and receiver can understand the academic paperwork being submitted. I am not a chemist, nor do I have the knowledge base to understand all the different types of pesticides that are used in the market. What was garnered is this the pesticide industry burst wide open in the 1940ies after World War II with knowledge from a poison gas. Modifying poison gas to be used on insects to get a greater yield from crops in the name of higher profits. The farming and agricultural industries have followed this business model of pesticides for 70 years without our consent. It is not a good or safe business model.

Action can be taken legislatively by states and the federal government. However, it is the average person that needs to become involved. Learn who your delegates are in the state legislature and who your house representative and senators are in Washington DC. Write your letters, it is easy to contact representatives by email and you do not need a stamp.

I learned farming can be done more efficiently as the Netherlands have accomplished. They have shown us in twenty years how to use a limited amount of land and became the second largest exporter of food in the world and using 95% less water. [384] Yes, the United States was almost beaten in food production by a country the size of Maryland.

The potential for transforming the farm-to-fork industry without the use of pesticides is great, and there are already examples of successful implementation of such practices. Vertical farming, for instance, involves growing crops in vertically stacked layers, using artificial light, and a controlled environment that eliminates the need for pesticides. This method uses less water and space than traditional farming, and it can be implemented in urban areas, reducing the transportation costs associated with long-distance shipping of produce.

Another promising approach is the use of agroecology, which focuses on promoting biodiversity, ecological processes, and local knowledge to enhance crop productivity, without relying on synthetic pesticides or fertilizers. This approach has been shown to be effective in increasing crop yields, reducing greenhouse gas emissions, and promoting the well-being of farmers and rural communities.

Additionally, the use of precision agriculture, which involves the use of sensors, robotics, and artificial intelligence to optimize crop production, has the potential to significantly reduce the use of pesticides. This approach allows farmers to monitor crop growth and apply fertilizers or pesticides only where they are needed, reducing waste, and minimizing the environmental impact.

---

384 https://www.youtube.com/watch?v=5clOYWsNhhk

In conclusion, there are many innovative methods that could transform the farm-to-fork industry, making it more sustainable, environmentally friendly, and healthy. With the help of visionary entrepreneurs and policymakers, we can create a new paradigm of farming that prioritizes the health of people and the planet over profit margins.

Precision fermentation is a rapidly growing technology that uses microorganisms to create animal-free products that mimic the taste and texture of traditional animal-based foods. This technology has the potential to revolutionize the food industry by providing sustainable and ethical alternatives to animal agriculture.

Perfect Day, a California-based animal-free dairy company, uses precision fermentation to produce a range of animal-free dairy products. Their process involves using microflora to produce casein and whey proteins, which are then combined with water, fats, and sugars to create a variety of dairy products such as milk, cheese, and ice cream. This process has numerous advantages over traditional animal agriculture, as it does not require the use of antibiotics, hormones, or animal feed, and has a significantly lower carbon footprint.

Similarly, Change Foods, another animal-free company, is also using precision fermentation to create dairy products that are free from animal agriculture. Their technology involves producing milk proteins using microbial fermentation, which are then combined with plant-based fats and sugars to create dairy products such as cheese and yogurt. This process has the potential to replace traditional animal agriculture by providing sustainable and ethical alternatives that are more environmentally friendly and humane.

These developments in precision fermentation have already attracted major players in the food industry, with Perfect Day already having agreements with companies such as Mars, Nestlé, Starbucks, and General Mills. Change Foods has also built a 1.2-million-liter custom-built fermentation facility in Abu Dhabi and plans to replace 10,000 dairy cows with this fermentation facility.

These companies are paving the way for a future of sustainable and ethical food production and may ultimately lead to the end of animal agriculture as we know it. [385]

Indeed, the way we grow food is changing, and there are many alternative solutions to feed the growing population without relying on harmful pesticides. One approach is to adopt organic farming methods, which eliminate the use of synthetic pesticides and rely on natural pest management techniques. Another approach is to use precision agriculture techniques, which rely on data and technology to optimize crop yields while reducing the use of pesticides and other chemicals.

Other innovative approaches to food production include aquaponics, which combines fish farming with hydroponic crop cultivation, and cellular agriculture, which involves growing animal products such as meat and dairy in a lab without the need for raising and slaughtering animals.

Overall, it is clear that there are many alternative solutions to feed everyone on our planet without relying on harmful pesticides. By embracing these solutions, we can create a healthier, more sustainable food system for all.

Closing credits

I would like to thank Therese Verner for assisting in editing *Without Our Consent,* Beth Weeks for the book cover design, and Brandon Stone for the narration. Please see the 399 footnotes in the eBook or the printed version. These footnotes document the thousands of hours of research which led to the creation of this book.

---

[385] https://www.gulfood.com/insights/dairy-desert-change-foods-bring-first-precision-fermentation-facility-uae

Appendix

Vitamin D added to milk

Milk has vitamin D added and it is essential for our health. Vitamin D helps with absorption of calcium and phosphorus in the small intestines. Without vitamin D, a deficiency can led to abnormalities in bone metabolism, such as rickets in children or osteoporosis (softening of the bones) in adults as well as heart disease and related complications. Older adults' deficiency in vitamin D can have some forms of dementia, including Alzheimer's disease. Research is linking low levels of vitamin D with diabetes. Other research is stating low blood levels of vitamin D and aggressive prostate cancer in European American and African American men. The Journal of Sexual Medicine found that men with severe erectile dysfunction (ED) had significantly lower vitamin D levels than men with mild ED. In the Basic and Clinical Research found vitamin D deficiency was associated with a higher breast cancer risk. [386]

THE FOLLOWING LIST REFLECTS MY PERSONAL RESULTS OF MY FOOD ALLERGIES DETERMINED BY TESTMYALLERGY.COM

---

[386] https://www.everydayhealth.com/news/illnesses-linked-vitamin-d-deficiency/

Butter
*A dairy product, made with the natural fat found
in milk (milk fat)* 99%

Spelt
A type of wheat, also known as Dinkel wheat. 99%

Milk lactose
*This indicates intolerance to lactose found
within dairy milk.* 98%

Beans (green)
Long, thin green in color 98%

Coffee (with sugar)
Hot drink with added sugar. 97%

Rice
Small white or brown grains 97%

Whisky
*A spirit distilled from malted grain, especially barley
or rye.* 97%

Buttermilk
*The liquid left behind after churning butter
out of cream.* 96%

Crayfish
Freshwater crustacean resembling a small lobster. 96%

Leather
*Material made from the skin of an animal by tanning
or other similar process.* 96%

Garlic
*A plant in the Allium (onion) family. It is closely
related to onions, shallots, and leeks.* 95%

Brazil nut
A large, three-sided South American nut 94%

Grapes (white)
This includes items made with grapes – wine. 92%

E 951
Aspartame Artificial sweetener 89%

Cotton

A soft white fibrous substance which surrounds the seeds of the cotton plant and is made into textile fiber and thread for sewing. 89%

Hop (Humulus lupulus)

A flowering plant – used for beer production 89%

Apples (Raw)

A fruit – numerous different species. Colors are usually green and red. 87%

Raspberries (raw)

Related to the blackberry, consisting of a cluster of reddish-pink drupelets. 87%

Apples (Cooked)

A fruit – numerous different species. Colors are usually green and red. 86%

Artichoke (Cooked)

A variety of thistle, cultivated for eating 86%

Lettuce Cultivated

Plant eaten in salads mostly. This includes all varieties of lettuce. 86%

Recipes

Lemon Caper Salmon
Use all organic ingredients.

1 tablespoon olive oil

¼ cup white wine

¼ cup white wine vinegar

¼ cup butter cut into cubes

1 tablespoon rosemary or favorite flavor parsley

2 tablespoons drained capers

1 tablespoon lemon juice

1 teaspoon grated ginger

Pinch of salt and pepper as desired.

Heat oven to 400 degrees.

Line baking sheet with foil (if available).

Brush or spray with olive oil.

Season both sides of salmon with rosemary.

Salt and pepper to taste.

In a saucepan add in wine, vinegar, butter.

Stir in capers and lemon.

Pour over salmon.

Home-made Wheat Bread
Use all organic ingredients.

1 to 1¼ cup of lukewarm water

¼ cup vegetable oil

¼ cup honey, or maple syrup

3½ cups wheat flour

2½ teaspoons yeast or 1 packet of active dry yeast.

Dissolve yeast in 2 tablespoons of water.

¼ cup almond milk

1 teaspoon sea salt

In a small bowl mix honey or maple syrup, and yeast in lukewarm water.

Let stand until creamy, about 10 minutes.

In mixing bowl mix all ingredients together to start the kneading process.

Let the dough rest in the bowl for about 20 to 30 minutes. Cover bowl with clean towel. This allows dough to absorb ingredients.

With mixing bowl or by hand, knead the dough for 6 to 8 minutes or until dough is supple. Dough will be soft, and firmly add water or flour if necessary.

Transfer dough to a lightly greased bowl and let rise for 1 hour to 2 hours depending on warmth of room.

After dough rises spray bread pan or grease bread pan and add teaspoon flour and shake flour to cover pan.

Transfer to bread to bread pan 8½" x 4½"

Let rise for 1 to 2 hours, depends on warmth of room.

Bake bread for 30 to 40 minutes, tent bread after 20 minutes with aluminum foil to prevent bread from getting too brown.

If using an instant read thermometer, bread will be 190 degrees Fahrenheit when done.

When complete, empty bread on bread rack to cool.

Add butter to crust if desired.

Coconut Chicken with Sundried Tomato
Use all organic ingredients.

2 organic chicken breasts

1 teaspoon organic oregano

1 teaspoon organic parsley

½ teaspoon salt or pinch of sea salt

¼ teaspoon ground or pinch organic black pepper

2 tablespoons of organic olive oil

1 cup chopped sweet organic onion

2 organic garlic cloves, minced

1 cup organic sundried tomato

2 tablespoons cooking organic white wine

2 tablespoons lemon juice

14½ oz. can of coconut milk

½ cup sliced fresh basil

In a small bowl add in all organic material minced garlic, oregano, parsley, onion powder, sea salt, and ground black pepper. Whip together.

Season each side of the chicken breasts with blend.

Add organic olive oil to a pan to accommodate two breast of chicken.

Medium heat: Cook each side for 5 to 10 minutes (depending on thickness) until they are turning white in color both sides. Best to use a cover while cooking.

In another pan, cook chopped onions with 2 tablespoons of organic olive oil until translucent, about 3 minutes.

Add minced garlic, stir in for about a minute.

Add in coconut milk and white wine (white cooking wine) I like to use chardonnay over cooking wine.

Stir in wine to fully incorporate the ingredients.

Add in the sundried tomatoes, and lemon juice.

Stir and bring to a simmer.

Add chicken to the cream sauce and let cook for 5 min.

Remove from heat, add basil and serve.

Chicken Basil and Feta for Two
Use all organic ingredients.

2 breast of chicken cut into strips

Salt and pepper for taste

¼ cup olive oil

3 tablespoon lemon juice

½ teaspoon thyme

½ teaspoon basil

¼ teaspoon onion salt

¾ cup feta

Heat oven to 350 degrees.

Cook chicken strips in olive oil about two minutes or until brown, chicken will finish cooking in the oven.

Place chicken in cooking dish in strips.

Mix lemon juice, basil add 2 more tablespoons olive oil, thyme, and onion salt.

Pour this mixture over chicken strips.

Add feta cheese on top of chicken strips.

Bake 10 to 15 minutes.

Oatmeal Cookies
Use all organic ingredients.

½ cup softened butter

¾ cup brown sugar

½ Turin sugar

2 large eggs

2 teaspoons vanilla

2¼ cup rolled oats

1¾ cups wheat flour

½ teaspoon baking soda

½ teaspoon sea salt

Preheat oven to 325 degrees.

Set aside parchment paper on cookie sheet.

Ia large bowl, Cream butter and sugars together until light and fluffy.

Mix for 3 or 4 minutes.

Stir in the eggs and vanilla until well combined.

Roll into small balls and place on parchment paper.

Bake for about 12 minutes.

Carrot Cake
Use all organic ingredients.

4 eggs

2 cups of grated carrots

½ of grated zucchini

2 cups wheat flour

½ cup Almond milk

¾ of a cup brown sugar

3 teaspoons of baking powder

¼ teaspoon ginger powder

a pinch of globe spice powder

a pinch of nutmeg

a pinch of sea salt

Mix eggs carrots, zucchini, almond milk, brown sugar, baking powder.

Mix powder and spices together, add in flour.

Beat for three or four minutes with mixer.

For the frosting:

one tablespoon vanilla extract

¾ of a cup cream cheese

¼ cup powdered sugar

Place in mixing bowl vanilla extract cream cheese and powdered sugar mixed till frothy.

Cook 30 min, oven 300 degrees. Yum!

Lemon Caper Salmon
Use all organic ingredients.

one tablespoon olive oil

salmon filets for two people

sea salt and pepper as needed

2 medium size shallots chopped (if desired)

¼ cup white wine vinegar

¼ cup white wine

½ organic butter slice in the cubes

2 tablespoons Rosemary

2 tablespoons drained capers

1 tablespoon of lemon juice

1 teaspoon grated ginger

Heat the oven to 400 degrees.

Spray organic olive oil on pan.

Season both sides of salmon with sea salt and pepper.

Bake for about 20 to 25 minutes or until salmon is flaky.

In a saucepan, simmer shallots in white wine vinegar until soft and very little liquid remains.

Add the butter and two tablespoons of water, stir constantly over the heat, until butter is melted and incorporated with the shallots.

Stir in parsley capers and lemon pour over the salmon after you pull it out of the oven.

Chicken Basil and Feta for 2
Use all organic ingredients.

2 chicken breasts cut into strips

¼ cup olive oil

3 lemons juiced

½ teaspoon basil

½ teaspoon thyme

¼ teaspoon onion salt

¾ Feta cheese crumbled up

Heat the oven to 350 degrees Fahrenheit.

Add salt and pepper, both sides of your chicken strips.

Add 2 tablespoons of olive oil to your pan and cook chicken strips until lightly brown, they'll finish cooking in the oven.

Lay the chicken strips on a piece napkin or something to soak up the oil.

Place the chicken strips in a baking dish.

Mix lemon juice and 2 tablespoons of olive oil, thyme, basil and onion salt until blended.

Pour this mixture over the chicken strips.

Sprinkle crumpled feta cheese over the chicken.

Bake for about 10 to 15 minutes.

These are a few of the recipes modified from the Internet to make sure I was eating organic. However, after several years of eating the same food over and over it really does get very old.

Buying organic flour
I purchased organic wheat flour from:
Azure Standard, 79709 Dufur Valley Road, Dufur, OR 97021
to make homemade bread and other deserts from organic flour.

ARTIFICIAL SWEETENERS

Artificial sweeteners, such as acesulfame potassium, are used in a variety of food and beverage products as a low-calorie sugar substitute. Annatto, a natural colorant, is often used in dairy products to enhance their appearance. Bleaching of dairy products is a process that uses hydrogen peroxide or other chemicals to remove the yellow color from milk or cream. Carmine, a red pigment derived from insects, is commonly used as a natural food coloring agent. Carrageenan is a plant-based ingredient that is often used as a thickening agent in food products, including dairy products.

Citric acid is a weak organic acid that is used as a preservative and flavoring agent in many food and beverage products. Monoglycerides and diglycerides are food additives that act as emulsifiers and stabilizers. They are often used in baked goods, dairy products, and margarine.

Sugars, such as fructose, glucose, or sucrose, are commonly added to food products to enhance their flavor and sweetness. Modified corn starch is a common ingredient in processed foods and is used to improve the texture and water retention properties of foods.

Polysorbate 80 is a food additive that is used as an emulsifier and stabilizer in a variety of food and beverage products. Cellulose is a plant-based ingredient that is often used to provide a creamy texture in food products.

Xanthan gum is a natural thickener that is commonly used in dairy products, salad dressings, and sauces. Lecithin is a food additive that is used as an emulsifier and is often found in baked goods and chocolate products.

Vitamin A acetate and vitamin A palmitate are commonly used as food additives to provide a source of vitamin A in processed foods. Calcium phosphate is often added to dairy products to improve their texture and to reduce the rennet coagulation time.

Disodium phosphate is a food additive that is commonly used as an emulsifier and to regulate acidity in processed foods. Guar gum is a plant-based ingredient that is commonly used as a thickener and stabilizer in food products.

Lactic acid is a natural organic acid that is often used as a preservative and flavoring agent in food products. Locust bean gum is a natural thickener that is often used in dairy products, sauces, and ice cream. Milk protein concentrate is a high-protein dairy ingredient that is often used in processed foods.

Natamycin is a natural antifungal agent that is often used as a food preservative to prevent the growth of yeasts and molds. Sodium phosphate is a food additive that is often used to extend the shelf life of food products by preventing spoilage. [387]

Not all of these items will be in the milk you buy but reading the label can tell you what is in that milk container.

FEDERAL LEGISLATIVE RULES FOR CORPORATIONS

Follow the rules with less than 500 employees. We need a better legislation to protect all living organisms.

- 50% fee waiver for businesses with < 500 employees and ≤ $60 M in global pesticides sales.

- 75% fee waiver for businesses with < 500 employees and ≤ $10 M in global pesticide sales.

---

[387] https://dontwastethecrumbs.com/30-additives-in-dairy-products-you-should-know-about/

FEE WAIVERS UNDER PRIA 1

Fee waivers for small businesses:

- 50% fee waiver for businesses with < 500 employees and ≤ $60 M in global pesticides sales.

- 75% fee waiver for businesses with < 500 employees and ≤ $10 M in global pesticide sales.

- 100% fee waivers to federal/state agencies and IR-4 submissions that meet criteria specified in FIFRA.

- Voluntary payment for registration action under review at the time of passage of PRIA 1.

PRIA 1 Funding for Worker Protection Activities

- $750,000 to 1 million per year. Funding for:

- Surveys on farm worker employment, health, living conditions, demographics.

- State oversight of farm worker illnesses and injuries.

- Training programs and materials for farm workers to reduce risks of pesticide use.

- Training for health care providers to recognize and treat pesticide poisonings.

Pesticide Registration Improvement Renewal Act (PRIA 2)

Congress reauthorized PRIA effective October 1, 2007, the beginning of Fiscal Year 2008. The Pesticide Registration Improvement Renewal Act (PRIA 2) authorized the fee system until September 30, 2012. PRIA 2:

- Expanded the number of fee categories of registration applications from 90 to 140.

- Continued funding for farm worker protection activities.

- Established funding for partnership grants and pesticide safety education programs.

Partnership Grants

- $500,000–750,000 per year.

- Administered under Pesticide Environmental Stewardship Program.

- Used in conjunction with appropriated funds to support innovative Integrated Pest Management projects that help to reduce pesticide risk.

Pesticide Registration Improvement Extension Act (PRIA 3)
Effective October 1, 2012, the Pesticide Registration Improvement Extension Act (PRIA 3) reauthorized the Pesticide Registration Improvement Renewal Act of 2007 (PRIA 2) for five years, until 2017. PRIA 3:

- Expanded number of fee categories of registration applications from 140 to 189.

- Inert clearances (I codes) included as covered PRIA categories.

- Miscellaneous categories included as covered PRIA categories.

- Established significant footnotes for covered PRIA categories.

- Continued funding for farm worker protection activities.

- Continued funding for partnership grants and pesticide safety education programs.

- Established funding for IT enhancements. (from maintenance fees)

- Continued 50% and 75% fee waiver reductions for small businesses.

- Continued exemption from requirement of fee for Federal/State agencies and for applications associated with IR-4 that meet certain criteria.

- Provided for maintenance fee reduction for certain small businesses.

- Process improvements in receiving PRIA application and for clear labels and label issue resolution period established.

Process Improvements
- 45/90 Preliminary Technical Screen.
- Clean label/Label Resolution Period (RD, AD only).
- IT enhancements.

Preliminary Technical Screen

The Preliminary Technical Screen allowed the Agency to identify deficiencies early in the application process and allows the applicant a 10-day period to correct these deficiencies at the front end of the review process.

- A 45-Day technical deficiency screen time frame for actions with decision review timeline of 6 months or less.

- A 90-day technical deficiency screen time frame for actions with decision review timeline of greater than 6 months.

- Deficient applications that are not corrected may be rejected, thus freeing up Agency resources for those applications that can proceed toward a regulatory decision.

These screens augment the 21-Day Completeness Screen established under PRIA 2, which checks to make sure all components of the application are present. This completeness screen does not include evaluation of quality of submission.

Clean Labels/Two-Day Label Review

PRIA 3 included a provision to ensure "clean" labels. Previously, some labels may have been approved by EPA with numerous conditions. The label conditions were placed on the accompanying label amendment, which made state enforcement of the labels more difficult. This applied to registration applications submitted under PRIA 3.

Label Resolution Period

The label resolution period applies to RD and AD actions only, and only to actions submitted under PRIA 3. EPA provided a draft accepted label to applicant on or before the PRIA due date. The applicant:

- Agreed to all of the terms associated with draft accepted label.

- Did not agree to one or more terms and requested additional time to resolve difference(s).

- Withdrew the application without prejudice for subsequent re-submission but forfeited the associated registration service fee.

An applicant who requested additional time to resolve differences had up to 30 days to reach agreement with the Agency on the final terms of the EPA-accepted label. If agreement was reached, EPA would have two business days after submission of the revised label to review and approve the final stamped label.

IT Enhancements

PRIA 3 provided funding for and required EPA to:

- Develop electronic tracking of registration submissions.

- Develop tracking of status of conditional registrations.

- Enhance the endangered species database.

- Allow for electronic submission and review of labels and confidential statements of formula.

Miscellaneous Changes

- PRIA 3 due dates that fell on a weekend or holiday were extended to the next business day.

- There was a limitation on the number of new product applications (5) with a new active ingredient or first food use application; each additional new product application was subject to its registration service fee, as was a new inert approval submitted with the new a.i. or first food use package.

- Applicant-initiated information submitted after completion of the technical deficiency screen was subject to 25% of the related fee for "new active ingredient and first food use" category.

- If an application was associated with and dependent upon a pending inert ingredient approval, the decision timeline for the associated application was extended to match the PRIA due date of the pending inert action, unless the due date for the associated action was further out, in which case it was subject to its own decision review timeline.

- There is a similar approach for applications associated with and dependent upon external review (Human Studies Review Board, Scientific Advisory Panel).

- If an inert approval application covers multiple inert ingredients grouped by EPA into one chemical class, a single registration service fee was assessed.

Pesticide Registration Improvement Extension Act of 2018 (PRIA 4)

Effective March 8, 2019, the Pesticide Registration Improvement Extension Act of 2018 (PRIA 3) reauthorized the Pesticide Registration Improvement Renewal Act (PRIA 3) for five years, through 2023. PRIA 4:

- Expands the number of covered categories from 189 to 212;

- Provides enhanced financial incentives for reduced-risk submissions;

- Continues small business fee waivers as well as exemption for federal and state agencies and IR-4 applications that meet certain criteria;

- Continues PRIA set-aside for farm worker protection activities, partnership grants and pesticide safety education programs;

- Raises the annual maintenance fee collection target from $27.8 million to $31 million;

- Eliminates appropriations constraint (the "1-to-1" provision) on spending FIFRA maintenance fees;

- Creates two new maintenance fee set-asides (up to $500,000/ year for each, from 2018 through 2023) for:

- efficacy guideline development and rulemaking for invertebrate pests of significant public health or economic importance;

- Good Laboratory Practices (GLP) inspection support;

- Eliminates maintenance fee set-aside for IT enhancements;

- Establishes new reporting requirements for (a) efficacy guidelines and rulemaking deliverables (b) GLP inspections (c) registration review mitigation implementation (d) identification of reforms to streamline new a.i. and new use review processes, (e) registration of pesticides to control vector-borne public health pests, and (e) EPA and stakeholder evaluation of the appropriateness and effectiveness of worker protection activities, partnership grants, and pesticide safety education programs funded under the PRIA set-aside. [388-]

MILKWEED YOU CAN PLANT BY REGION

Common Milkweed (Asclepias syriaca)

Native Range: AL, AR, CT, DC, DE, GA, IA, IL, IN, KS, KY, LA, MA, MD, ME, MI, MN, MO, MS, MT, NC, ND, NE, NH, NJ, NY, OH, OK, OR, PA, RI, SC, SD, TN, TX, VA, VT, WI, WV. See range map. Description: This tall perennial has large balls of pink or purplish flowers that have an attractive odor. The flowers bloom from June to August.

Growing Conditions: Shade intolerant, needs lots of sunlight, moist soil.

Plant Size: Usually 3-5 feet (90-150 cm), sometimes reaching 8 feet (240 cm) in ditches and gardens.

Butterflyweed (Asclepias tuberosa)

Native Range: AL, AR, AZ, CA, CO, CT, DC, DE, FL, GA, IA, IL, IN, KS, KY, LA, MA, MD, ME, MI, MN, MO, MS, NC, NE, NH, NJ, NM, NY, OH, OK, PA, RI, SC, SD, TN, TX, UT, VA, VT, WI, WV

Description: Sometimes called orange milkweed, this perennial has large, flat-topped clusters of yellow-orange or bright-orange

---

388 https://www.epa.gov/pria-fees/pria-overview-and-history

flowers and blooms May to September.

Growing Conditions: Needs sunlight, drought tolerant, dry or moist soil.

Plant Size: 1-2 ft (30-60 cm)

Swamp Milkweed (Asclepias incarnata)

Native Range: AL, AR, CO, CT, DC, DE, FL, GA, IA, ID, IL, IN, KS, KY, LA, MA, MD, ME, MI, MN, MO, MT, NC, ND, NE, NH, NJ, NM, NV, NY, OH, OK, PA, RI, SC, SD, TN, TX, UT, VA, VT, WI, WV, WY

Description: Also known as pink milkweed, this perennial has large blossoms composed of small, rose-purple flowers. The deep pink flowers are clustered at the top of a tall, branching stem and bloom June to October.

Growing Conditions: Needs lots of water, shade tolerant, moist to wet soil.

Plant Size: 2-5 ft (60-152 cm)

Antelope-horns Milkweed (Asclepias asperula)

Native Range: AZ, CA, CO, ID, KS, NE, NM, NV, OK, TX, UT

Description: Also known as spider milkweed, this perennial is clump-forming with stems that are densely covered with minute hairs. As the green seed pods grow, they curve to resemble antelope horns. It has pale, greenish-yellow flowers, tinged maroon that bloom March to October.

Growing Conditions: Needs sunlight, dry or moist soil, medium water use.

Plant Size: 1-2 ft (30-60 cm) tall

Purple Milkweed (Asclepias purpurascens)

Native Range: AR, CT, DC, DE, GA, IA, IL, IN, KS, KY, LA, MA, MD, MI, MN, MO, MS, NC, NE, NH, NJ, NY, OH, OK, PA, RI, SD, TN, TX, VA, WI, WV

Description: The flowers of this beautiful milkweed species are rich magenta-purple and bloom May to July.

Growing Conditions: Needs sunlight and dry soil.

Plant Size: 2-4 ft (61 to 122 cm)

Showy Milkweed (Asclepias speciosa)

Native Range: AZ, CA, CO, IA, ID, IL, KS, MI, MN, MT, ND, NE, NM, NV, OK, OR, SD, TX, UT, WA, WI, WY

Description: This perennial has large, oval, blue-green leaves and spherical clusters of rose-colored flowers. The flowers occur at the top of the stem and on stalks from leaf axils and bloom May to September.

Growing Conditions: Shade intolerant, needs sunlight, medium water use, moist soil.

Plant Size: Generally, 1 ½ – 3 ft (46 – 91 cm) but can reach 6 ft (183 cm) under favorable conditions.

California Milkweed (Asclepias californica)

Native Range: Central and southern California

Description: This perennial is a white-woolly plant with milky sap and deep purple flowers. It blooms from May to July.

Growing Conditions: Drought tolerant, dry slopes

Plant Size: Maximum height 3 ft (91 cm)

White milkweed (Asclepias variegata)

Native Range: AL, AR, CT, DC, DE, FL, GA, IL, IN, KY, LA, MD, MO, MS, NC, NJ, NY, OH, OK, PA, SC, TN, TX, VA, WV

Description: This perennial has small white flowers with purplish centers crowded into round, terminal clusters that resemble snowballs and blooms from May to September.

Growing Conditions: Low water use, dry soil, moderately shade tolerant.

Plant Size: 1-3 ft (30- 91 cm)

Whorled milkweed (Asclepias verticillata)

Native Range: AL, AR, AZ, CT, DC, DE, FL, GA, IA, IL, IN, KS, KY, LA, MA, MD, MI, MN, MO, MS, MT, NC, ND, NE, NJ, NM, NY, OH, OK, PA, RI, SC, SD, TN, TX, VA, VT, WI, WV, WY

Description: This single-stemmed perennial has narrow, linear leaves whorled along the stem. Small, greenish-white flowers occur in flat-topped clusters on the upper part of the stem and bloom May to September.

Growing Conditions: Low water use, moderately shade tolerant, dry soil.

Plant Size: 1-3 ft (30- 91 cm)

Mexican Whorled Milkweed (Asclepias fascicularis)

Native Range: CA, ID, NV, OR, UT, WA

Description: Also known as narrowleaf milkweed, this perennial has narrow, whorled leaves with clusters of greenish-white flowers, often tinged with purple and blooms June to September.

Growing Conditions: Needs sunlight, drought tolerant, dry to moist soils.

Plant Size: 1-2½ ft (30-76 cm)

Desert Milkweed (Asclepias erosa)

Native Range: AZ, CA, NV, UT

Description: Desert milkweed has white to yellow flowers and a green to yellow stem and blooms April to October. Identification is somewhat difficult because its leaves vary from mostly smooth to covered with fine cream-colored hair.

Growing Conditions: Best grown in deserts or desert conditions with sandy soils, needs sunlight, dry soils, not shade tolerant.

Plant Size: 1-3 ft (30- 91 cm)

Green Milkweed (Asclepias viridis)

Native Range: AL, AR, FL, GA, IL, IN, KS, KY, LA, MO, MS, NE, OH, OK, SC, TN, TX, WV

Description: Also known as green antelopehorn milkweed, this perennial has white flowers – mostly one per plant and lacks the "horns" seen on antelopehorn milkweed. These milkweeds bloom from May to August.

Growing Conditions: Needs sunlight, cold and heat tolerant, moist soil, low water use.

Plant Size: Matures to 4 ft (122 cm) in height.

Garden Habits native milkweed for monarchs Abby Barber & David Mizejewshi [389]

USDA ORGANIC REGULATIONS

From the USDA these are rules for organic non-organic pesticides

'In organic crop production, non-synthetic (natural) substances are allowed unless specifically prohibited and synthetic substances are prohibited unless specifically allowed (§§ 205.601 - 205.602)

In organic livestock production, nonsynthetic (natural) substances are allowed unless specifically prohibited and synthetic substances are prohibited unless specifically allowed (§§ 205.603 - 205.604) [390]

"§ 205.601 Synthetic substances allowed for use in organic crop production.

In accordance with restrictions specified in this section, the following synthetic substances may be used in organic crop production: Provided, That, use of such substances do not contribute to contamination of crops, soil, or water. Substances allowed by this section, except disinfectants and sanitizers in paragraph (a) and those substances in paragraphs (c), (j), (k), (l), and (o) of

---

389 https://blog.nwf.org/2015/02/twelve-native-milkweeds-for-monarchs/
390 https://www.ams.usda.gov/rules-regulations/organic/national-list

this section, may only be used when the provisions set forth in § 205.206(a) through (d) prove insufficient to prevent or control the target pest.

(a) As algicide, disinfectants, and sanitizer, including irrigation system cleaning systems.

(1) Alcohols.

(i) Ethanol.

(ii) Isopropanol.

(2) Chlorine materials – For pre-harvest use, residual chlorine levels in the water in direct crop contact or as water from cleaning irrigation systems applied to soil must not exceed the maximum residual disinfectant limit under the Safe Drinking Water Act, except that chlorine products may be used in edible sprout production according to EPA label directions.

(i) Calcium hypochlorite.

(ii) Chlorine dioxide.

(iii) Hypochlorous acid – generated from electrolyzed water.

(iv) Potassium hypochlorite – for use in water for irrigation purposes.

(v) Sodium hypochlorite.

(3) Copper sulfate – for use as an algicide in aquatic rice systems, is limited to one application per field during any 24-month period. Application rates are limited to those which do not increase baseline soil test values for copper over a timeframe agreed upon by the producer and accredited certifying agent.

(4) Hydrogen peroxide.

(5) Ozone gas – for use as an irrigation system cleaner only.

(6) Peracetic acid – for use in disinfecting equipment, seed, and asexually propagated planting material. Also permitted in hydrogen peroxide formulations as allowed in § 205.601(a) at concentration of no more than 6% as indicated on the pesticide product label.

(7) Soap-based algicide/demossers.

(8) Sodium carbonate peroxyhydrate (CAS #-15630-89-4) – Federal law restricts the use of this substance in food crop production to approved food uses identified on the product label.

(b) As herbicides, weed barriers, as applicable.

(1) Herbicides, soap-based - for use in farmstead maintenance (roadways, ditches, right of ways, building perimeters) and ornamental crops.

(2) Mulches.

(i) Newspaper or other recycled paper, without glossy or colored inks.

(ii) Plastic mulch and covers (petroleum-based other than polyvinyl chloride (PVC)).

(iii) Biodegradable biobased mulch film as defined in § 205.2. Must be produced without organisms or feedstock derived from excluded methods.

(c) As compost feedstocks – Newspapers or other recycled paper, without glossy or colored inks.

(d) As animal repellents – Soaps, ammonium – for use as a large animal repellant only, no contact with soil or edible portion of crop.

(e) As insecticides (including acaricides or mite control).

(1) Ammonium carbonate – for use as bait in insect traps only, no direct contact with crop or soil.

(2) Aqueous potassium silicate (CAS #-1312-76-1) – the silica, used in the manufacture of potassium silicate, must be sourced from naturally occurring sand.

(3) Boric acid – structural pest control, no direct contact with organic food or crops.

(4) Copper sulfate – for use as tadpole shrimp control in aquatic rice production, is limited to one application per field during any 24-month period. Application rates are limited to levels which do

not increase baseline soil test values for copper over a timeframe agreed upon by the producer and accredited certifying agent.

(5) Elemental sulfur.

(6) Lime sulfur – including calcium polysulfide.

(7) Oils, horticultural – narrow range oils as dormant, suffocating, and summer oils.

(8) Soaps, insecticidal.

(9) Sticky traps/barriers.

(10) Sucrose octanoate esters (CAS #s - 42922-74-7; 58064-47-4) – in accordance with approved labeling.

(f) As insect management. Pheromones.

(g) As rodenticides. Vitamin D_3.

(h) As slug or snail bait.

(1) Ferric phosphate (CAS # 10045-86-0).

(2) Elemental sulfur.

(i) As plant disease control.

(1) Aqueous potassium silicate (CAS #-1312-76-1) – the silica, used in the manufacture of potassium silicate, must be sourced from naturally occurring sand.

(2) Coppers, fixed – copper hydroxide, copper oxide, copper oxychloride, includes products exempted from EPA tolerance, *Provided,* That, copper-based materials must be used in a manner that minimizes accumulation in the soil and shall not be used as herbicides.

(3) Copper sulfate – Substance must be used in a manner that minimizes accumulation of copper in the soil.

(4) Hydrated lime.

(5) Hydrogen peroxide.

(6) Lime sulfur.

(7) Oils, horticultural, narrow range oils as dormant, suffocating,

and summer oils.

(8) Peracetic acid – for use to control fire blight bacteria. Also permitted in hydrogen peroxide formulations as allowed in § 205.601(i) at concentration of no more than 6% as indicated on the pesticide product label.

(9) Potassium bicarbonate.

(10) Elemental sulfur.

(11) Polyoxin D zinc salt.

(j) As plant or soil amendments.

(1) Aquatic plant extracts (other than hydrolyzed) – Extraction process is limited to the use of potassium hydroxide or sodium hydroxide; solvent amount used is limited to that amount necessary for extraction.

(2) Elemental sulfur.

(3) Humic acids – naturally occurring deposits, water and alkali extracts only.

(4) Lignin sulfonate - chelating agent, dust suppressant.

(5) Magnesium oxide (CAS # 1309-48-4) – for use only to control the viscosity of a clay suspension agent for humates.

(6) Magnesium sulfate – allowed with a documented soil deficiency.

(7) Micronutrients – not to be used as a defoliant, herbicide, or desiccant. Those made from nitrates or chlorides are not allowed. Micronutrient deficiency must be documented by soil or tissue testing or other documented and verifiable method as approved by the certifying agent.

(i) Soluble boron products.

(ii) Sulfates, carbonates, oxides, or silicates of zinc, copper, iron, manganese, molybdenum, selenium, and cobalt.

(8) Liquid fish products – can be pH adjusted with sulfuric, citric or phosphoric acid. The amount of acid used shall not exceed the minimum needed to lower the pH to 3.5.

(9) Vitamins, C and E.

(10) Squid byproducts – from food waste processing only. Can be pH adjusted with sulfuric, citric, or phosphoric acid. The amount of acid used shall not exceed the minimum needed to lower the pH to 3.5.

(11) Sulfurous acid (CAS # 7782-99-2) for on-farm generation of substance utilizing 99% purity elemental sulfur per paragraph (j) (2) of this section.

(k) As plant growth regulators.

(1) Ethylene gas – for regulation of pineapple flowering.

(2) Fatty alcohols (C6, C8, C10, and/or C12) - for sucker control in organic tobacco production.

(l) As floating agents in postharvest handling. Sodium silicate - for tree fruit and fiber processing.

(m) As synthetic inert ingredients as classified by the Environmental Protection Agency (EPA), for use with non § 205.601 Synthetic substances allowed for use in organic crop production.

In accordance with restrictions specified in this section, the following synthetic substances may be used in organic crop production: Provided, That, use of such substances do not contribute to contamination of crops, soil, or water. Substances allowed by this section, except disinfectants and sanitizers in paragraph (a) and those substances in paragraphs (c), (j), (k), (l), and (o) of this section, may only be used when the provisions set forth in § 205.206(a) through (d) prove insufficient to prevent or control the target pest.

(a) As algicide, disinfectants, and sanitizer, including irrigation system cleaning systems.

(1) Alcohols.

(i) Ethanol.

(ii) Isopropanol.

(2) Chlorine materials – For pre-harvest use, residual chlorine levels in the water in direct crop contact or as water from cleaning irrigation systems applied to soil must not exceed the maximum residual disinfectant limit under the Safe Drinking Water Act, except that chlorine products may be used in edible sprout production according to EPA label directions.

(i) Calcium hypochlorite.

(ii) Chlorine dioxide.

(iii) Hypochlorous acid – generated from electrolyzed water.

(iv) Potassium hypochlorite - for use in water for irrigation purposes.

(v) Sodium hypochlorite.

(3) Copper sulfate –- for use as an algicide in aquatic rice systems, is limited to one application per field during any 24-month period. Application rates are limited to those which do not increase baseline soil test values for copper over a timeframe agreed upon by the producer and accredited certifying agent.

(4) Hydrogen peroxide.

(5) Ozone gas – for use as an irrigation system cleaner only.

(6) Peracetic acid – for use in disinfecting equipment, seed, and asexually propagated planting material. Also permitted in hydrogen peroxide formulations as allowed in § 205.601(a) at concentration of no more than 6% as indicated on the pesticide product label.

(7) Soap-based algicide/demossers.

(8) Sodium carbonate peroxyhydrate (CAS #-15630-89-4) – Federal law restricts the use of this substance in food crop production to approved food uses identified on the product label.

(b) As herbicides, weed barriers, as applicable.

(1) Herbicides, soap-based – for use in farmstead maintenance (roadways, ditches, right of ways, building perimeters) and ornamental crops.

(2) Mulches.

(i) Newspaper or other recycled paper, without glossy or colored inks.

(ii) Plastic mulch and covers (petroleum-based other than polyvinyl chloride (PVC)).

(iii) Biodegradable biobased mulch film as defined in § 205.2. Must be produced without organisms or feedstock derived from excluded methods.

(c) As compost feedstocks – Newspapers or other recycled paper, without glossy or colored inks.

(d) As animal repellents – Soaps, ammonium - for use as a large animal repellant only, no contact with soil or edible portion of crop.

(e) As insecticides (including acaricides or mite control).

(1) Ammonium carbonate – for use as bait in insect traps only, no direct contact with crop or soil.

(2) Aqueous potassium silicate (CAS #-1312-76-1) – the silica, used in the manufacture of potassium silicate, must be sourced from naturally occurring sand.

(3) Boric acid – structural pest control, no direct contact with organic food or crops.

(4) Copper sulfate – for use as tadpole shrimp control in aquatic rice production, is limited to one application per field during any 24-month period. Application rates are limited to levels which do not increase baseline soil test values for copper over a timeframe agreed upon by the producer and accredited certifying agent.

(5) Elemental sulfur.

(6) Lime sulfur – including calcium polysulfide.

(7) Oils, horticultural – narrow range oils as dormant, suffocating, and summer oils.

(8) Soaps, insecticidal.

(9) Sticky traps/barriers.

(10) Sucrose octanoate esters (CAS #s - 42922-74-7; 58064-47-4) - in accordance with approved labeling.

(f) As insect management. Pheromones.

(g) As rodenticides. Vitamin D3.

(h) As slug or snail bait.

(1) Ferric phosphate (CAS # 10045-86-0).

(2) Elemental sulfur.

(i) As plant disease control.

(1) Aqueous potassium silicate (CAS #-1312-76-1) – the silica, used in the manufacture of potassium silicate, must be sourced from naturally occurring sand.

(2) Coppers, fixed – copper hydroxide, copper oxide, copper oxychloride, includes products exempted from EPA tolerance, Provided, That, copper-based materials must be used in a manner that minimizes accumulation in the soil and shall not be used as herbicides.

(3) Copper sulfate – Substance must be used in a manner that minimizes accumulation of copper in the soil.

(4) Hydrated lime.

(5) Hydrogen peroxide.

(6) Lime sulfur.

(7) Oils, horticultural, narrow range oils as dormant, suffocating, and summer oils.

(8) Peracetic acid – for use to control fire blight bacteria. Also permitted in hydrogen peroxide formulations as allowed in § 205.601(i) at concentration of no more than 6% as indicated on

the pesticide product label.

(9) Potassium bicarbonate.

(10) Elemental sulfur.

(11) Polyoxin D zinc salt.

(j) As plant or soil amendments.

(1) Aquatic plant extracts (other than hydrolyzed) - Extraction process is limited to the use of potassium hydroxide or sodium hydroxide; solvent amount used is limited to that amount necessary for extraction.

(2) Elemental sulfur.

(3) Humic acids – naturally occurring deposits, water and alkali extracts only.

(4) Lignin sulfonate – chelating agent, dust suppressant.

(5) Magnesium oxide (CAS # 1309-48-4) – for use only to control the viscosity of a clay suspension agent for humates.

(6) Magnesium sulfate – allowed with a documented soil deficiency.

(7) Micronutrients – not to be used as a defoliant, herbicide, or desiccant. Those made from nitrates or chlorides are not allowed. Micronutrient deficiency must be documented by soil or tissue testing or other documented and verifiable method as approved by the certifying agent.

(i) Soluble boron products.

(ii) Sulfates, carbonates, oxides, or silicates of zinc, copper, iron, manganese, molybdenum, selenium, and cobalt.

(8) Liquid fish products – can be pH adjusted with sulfuric, citric or phosphoric acid. The amount of acid used shall not exceed the minimum needed to lower the pH to 3.5.

(9) Vitamins, C and E.

(10) Squid byproducts – from food waste processing only. Can be pH adjusted with sulfuric, citric, or phosphoric acid. The amount of

acid used shall not exceed the minimum needed to lower the pH to 3.5.

(11) Sulfurous acid (CAS # 7782-99-2) for on-farm generation of substance utilizing 99% purity elemental sulfur per paragraph (j) (2) of this section.

(k) As plant growth regulators.

(1) Ethylene gas – for regulation of pineapple flowering.

(2) Fatty alcohols (C6, C8, C10, and/or C12) – for sucker control in organic tobacco production.

(l) As floating agents in postharvest handling. Sodium silicate – for tree fruit and fiber processing.

(m) As synthetic inert ingredients as classified by the Environmental Protection Agency (EPA), for use with nonsynthetic substances or synthetic substances listed in this section and used as an active pesticide ingredient in accordance with any limitations on the use of such substances.

(1) EPA List 4 – Inerts of Minimal Concern.

(2) EPA List 3 – Inerts of unknown toxicity – for use only in passive pheromone dispensers.

(n) Seed preparations. Hydrogen chloride (CAS # 7647-01-0) – for delinting cotton seed for planting.

(o) Production aids.

(1) Microcrystalline cheesewax (CAS #'s 64742-42-3, 8009-03-08, and 8002-74-2) - for use in log grown mushroom production. Must be made without either ethylene-propylene co-polymer or synthetic colors.

(2) Paper-based crop planting aids as defined in § 205.2. Virgin or recycled paper without glossy paper or colored inks.

(p)-(z) [Reserved] synthetic substances or synthetic substances listed in this section and used as an active pesticide ingredient in accordance with any limitations on the use of such substances.

(1) EPA List 4 – Inerts of Minimal Concern.

(2) EPA List 3 – Inerts of unknown toxicity - for use only in passive pheromone dispensers.

(n) Seed preparations. Hydrogen chloride (CAS # 7647-01-0) – for delinting cotton seed for planting.

(o) Production aids.

(1) Microcrystalline cheesewax (CAS #'s 64742-42-3, 8009-03-08, and 8002-74-2) – for use in log grown mushroom production. Must be made without either ethylene-propylene co-polymer or synthetic colors.

(2) Paper-based crop planting aids as defined in § 205.2. Virgin or recycled paper without glossy paper or colored inks.

(p)-(z) [Reserved]" [391]

PESTICIDES RESIDUE BURDEN SCORE

In this study, researchers used a pesticide residue burden score (PRBS) to assess dietary exposure to pesticides over the past year. The PRBS was based on a food frequency questionnaire and surveillance data on food pesticide residues. The study aimed to evaluate the association between the PRBS and urinary concentrations of pesticide biomarkers.

The study included 90 men from the EARTH study, and their fruit and vegetable intake was classified as having high (PRBS ≥ 4) or low (PRBS < 4) pesticide residues. The researchers analyzed two urine samples per man for seven biomarkers of organophosphate and pyrethroid insecticides and the herbicide 2,4-dichlorophenoxyacetic acid.

The results showed that men with high PRBS scores had significantly higher urinary concentrations of pesticide biomarkers compared to men with low PRBS scores. Specifically, men with high PRBS scores had higher concentrations of metabolites of organophosphate insecticides and the herbicide 2,4-dichlorophe

---

391 https://www.ecfr.gov/current/title-7/subtitle-B/chapter-I/subchapter-M/part-205/subpart-G/subject-group-ECFR0ebc5d139b750cd

noxyacetic acid. These findings suggest that the PRBS can be a useful tool in assessing dietary exposure to pesticides and predicting urinary biomarker concentrations. The study highlights the importance of reducing pesticide use in agriculture and increasing the availability of pesticide-free foods to minimize the potential health risks associated with exposure to pesticides.[392]

Polychlorinated Biphenyls (PCBs) are lipophilic chemicals first produced in 1929 and are commonly used in electronics manufacture. [393] In addition, PCBs, phytoestrogens, fungicides, pesticides, and other xenobiotics can disrupt brain sexual differentiation. This type of disruption has a high likelihood of affecting both reproductive physiology and behavior later in life, and indeed there is strong evidence in rodent models that reproductive success is diminished as a consequence. [394]

Enlist E3 soybeans

The development of a new type of soybean seed, Enlist E3 soybeans, which is a result of a collaboration between Dow AgroSciences and MS Technologies. This new soybean seed will revolutionize weed control and yield performance in soybeans.

The Enlist E3 trait is a molecular stack that provides exceptional tolerance to herbicides, including Enlist Duo™ herbicide, which is a combination of 2,4-D choline and glyphosate, and glufosinate herbicides. This robust tolerance to herbicides will enable growers to meet the ever-increasing weed challenges they face while also maximizing per-acre profits.

This development of Enlist E3 soybeans is a significant milestone in the agricultural industry. The combination of advanced genetics and herbicide tolerance technology will result in better weed control, higher yields, and more profitable farming operations.

Enlist E3 trait will enable farmers to use a wider range of herbicides in their weed management strategies, thereby providing

---

392 https://www.ncbi.nlm.nih.gov/pmc/articles/PMC5734986/
393 https://www.epa.gov/pcbs/learn-about-polychlorinated-biphenyls-pcbs
394 https://www.ncbi.nlm.nih.gov/pmc/articles/PMC2726844/

more flexibility in their farming practices. This flexibility will enable them to choose the most effective herbicide for the specific weed challenge they face while minimizing the potential for herbicide resistance.

In addition, Enlist E3 trait will help to reduce the environmental impact of herbicide use in agriculture. By providing farmers with better weed control options, they will be able to reduce the amount of herbicide required, leading to a more sustainable agricultural industry. Enlist E3 soybeans is a significant STEP forward in the agricultural industry, providing farmers with better weed control options, higher yields, and more profitable farming operations. [395]

In addition, Dow AgroSciences LLC and M.S. Technologies LLC have collaborated to develop a new soybean seed known as "DAS-444Ø6-6 soybean plants". This seed has been genetically modified to express three different proteins: aryloxyalkanoate dioxygenase-12 (AAD-12), double mutant 5-enolpyruvylshikimate-3-phosphate synthase (2mEPSPS), and phosphinothricin acetyltransferase (PAT).

The AAD-12 protein is an enzyme that has alpha-ketoglutarate-dependent dioxygenase activity. This results in the metabolic inactivation of the herbicides belonging to the aryloxyalkanoate family. The 2mEPSPS protein, on the other hand, provides resistance to glyphosate-based herbicides such as Roundup. Finally, the PAT protein provides resistance to glufosinate herbicides.

The development of these three proteins allows the soybean plant to resist the harmful effects of certain herbicides while continuing to grow and produce a high yield. This is an important development for farmers, as they can now effectively manage weed control and maximize their profits per acre. However, there are also concerns regarding the potential environmental impact of genetically modified crops, as well as the potential risks to human health. [396]

395 https://www.enlist.com/content/hdas/en-CA/how-it-works/enlist-traits.html
396 https://www.aphis.usda.gov/brs/aphisdocs/11_23401p.pdf

DICHLOROPHENOL

2,4-Dichlorophenoxyacetic acid (2,4-D), a phenoxyalkanoic acid herbicide, is among the most widely distributed pollutants in the environment. 2,4-Dichlorophenol (2,4-DCP), as the main metabolite of 2,4-D, always accompanies 2,4-D. In this paper, we did research on the combined toxicities of 2,4-D and 2,4-DCP to Vibrio qinghaiensis sp.-Q67 (Q67) and Caenorhabditis elegans. It was found that the toxicity of 2,4-DCP is more severe than that of its parent 2,4-D at any concentration levels whether to Q67 or to C. elegans. Furthermore, 2,4-DCP to Q67 has the time-dependent toxicity. The toxicity of the mixture of 2,4-D and 2,4-DCP to Q67 is increasing with the exposure time, but that to C. elegans does not change over time. There is a good linear relationship between the pEC50/pLC50 value of binary mixture ray of 2,4-D and 2,4-DCP and the mixture ratio of 2,4-DCP, which implies the predictability of mixture toxicity of 2,4-D and 2,4-DCP. The toxicological interactions of the binary mixtures to Q67 are basically additive actions whether at 0.25 or at 12 h. However, most mixtures have antagonistic interactions against C. elegans. [397]

NEWEST PESTICIDES BANNED LIST [398]

Anilazine	Barban
Benalaxyl	Benodanil
Bioallethrin /esbiothrin	Bromofenoxim
Bromoxynil butyrate	Bronopol
Copper oxide	Cycloate
Cyclosulfamuron	Cypermethrin, alpha
Cypermethrin, theta	Cypermethrin, zeta
Cyproconazole	Dichlofluanid
Dichlone	Difenzoquat
Dimethipin	Diquat
Disodium methanearsonate	(DSMA)

---

397 https://pubs.acs.org/doi/10.1021/acsomega.8b02282
398 https://www.pesticideinfo.org/resources/banned-pesticides

TOP 20 CHEMICAL COMPANIES

There are an overabundance of chemical companies serving the farming and livestock industry. The top 20 agrobusiness giants accounted for 95% of total sales in the world and the top 5 companies accounted for 54.71% of the total sales. (Syngenta, Bayer, CropScience, BASF and Corteva.) Below are the top twenty companies.

Syngenta Corp: A Swiss multinational company that produces agrochemicals, seeds, and crop protection products. Syngenta is a subsidiary of ChemChina.

Bayer Crop Science: A German company that produces pesticides, seeds, and herbicides. Bayer Crop Science is a subsidiary of the Bayer AG group.

BASF: A German chemical company that produces a wide range of products including agrochemicals and seeds.

Corteva: An American company that produces agricultural products including seeds, crop protection products, and digital tools to improve farming practices.

UPL: An Indian company that produces crop protection products, seeds, and specialty chemicals.

FMC: An American company that produces crop protection products, including insecticides, herbicides, and fungicides.

ADAMA: An Israeli company that produces crop protection products, including herbicides, fungicides, and insecticides.

Sumitomo Chemical: A Japanese company that produces a wide range of products including crop protection products and seeds.

Nufarm: An Australian company that produces herbicides, insecticides, and fungicides.

Dow AgroSciences: An American company that produces seeds and crop protection products, including herbicides, fungicides, and insecticides.

Monsanto: An American company that produces seeds and crop

protection products, including herbicides and genetically modified organisms (GMOs).

DuPont: An American company that produces seeds and crop protection products, including herbicides and insecticides.

Arysta LifeScience: A Japanese company that produces crop protection products, including herbicides, fungicides, and insecticides.

ADAMA Agricultural Solutions: An Israeli company that produces crop protection products, including herbicides, fungicides, and insecticides.

Nippon Soda: A Japanese company that produces crop protection products, including herbicides and insecticides.

Cheminova: A Danish company that produces crop protection products, including herbicides, fungicides, and insecticides.

Kumiai Chemical Industry: A Japanese company that produces crop protection products, including herbicides and insecticides.

Isagro: An Italian company that produces crop protection products, including fungicides and insecticides.

Gowan Company: An American company that produces crop protection products, including herbicides, fungicides, and insecticides.

Makhteshim Agan: An Israeli company that produces crop protection products, including herbicides, fungicides, and insecticides.

Seeds [399]

CONTACT PAN – Pesticide Action Network

Main Office: 2029 University Ave., Suite 200,
Berkeley, CA 94704

Phone: 510.788.9020

---

399 https://www.worldbenchmarkingalliance.org/publication/access-to-seeds-index/findings/companies-are-still-only-concentrating-their-investments-in-infrastructure-in-a-few-countries/

PAN'S BOARD OF DIRECTOR

Associate Professor, Kamakakūokalani

Malia Akutagawa is an Associate Professor of Law and Hawaiian Studies with both the University of Hawai'i at Mānoa's Hawai'inuiākea School of Hawaiian Knowledge – Kamakakūokalani Center for Hawaiian Studies and the William S. Richardson School of Law. Malia earned Baccalaureate degrees in Philosophy and Biology from Whitworth University in 1993. Malia is a 1997 alumnus of the William S. Richardson School of Law, having earned a Juris Doctor and Environmental Law Certificate. She was admitted into the Hawai'i State Bar Association in 1998.

Malia's scholarship includes State and federal laws protecting iwi kūpuna (Native ancestral burials), preserving cultural and historic sites, and engaging Native communities and stakeholders in consultation on these matters. Malia is also involved in community-based resource management efforts along traditional land divisions (ahupua'a) and district/regional (moku) levels within the context of State watershed management partnerships and Community Based Subsistence Fishing Areas (CBSFAs). Malia is particularly interested in the integration of Native, Indigenous Hawaiian methodologies, customary law, and governance principles founded by the ancient 'Aha Kiole (People's Councils) and incorporated into law under the Statewide'Aha Moku Advisory Committee (AMAC). The AMAC is comprised of representatives from each of the eight main Hawaiian islands. It advises the State Department of Land and Natural Resources (DLNR) on Indigenous best practices for the management of natural resources in Hawai'i.

Malia is part of Hui 'Āina Momona, a consortium of scholars throughout the university community charged with addressing compelling issues of indigenous Hawaiian knowledge and practices, including the legal regime and Native Hawaiian rights associated with mālama 'āina, and with focus on cross-disciplinary solutions to natural and cultural resource management, sustainability, and food security.

Office Phone: 808-956-0559

Office Fax: 808-973-0988

Office Location: KAMA 103DD

E-mail: maliaaku@hawaii.edu

Index